Remanufacturing and Advanced Machining Processes for New Materials and Components

Remanufacturing and Advanced Machining Processes for New Materials and Components

E. S. Gevorkyan, M. Rucki, V. P. Nerubatskyi,
W. Żurowski, Z. Siemiątkowski, D. Morozow,
and A. G. Kharatyan

First edition published 2022
by Taylor & Francis
6000 Broken Sound Parkway NW, Suite 300, Boca Raton, FL 33487-2742

and by Taylor & Francis
2 Park Square, Milton Park, Abingdon, Oxon, OX14 4RN

The Open Access option was financed by Polish Ministry of Education and Science in the frame of program 'Excellent Science' project No. DNM/SP/512333/2021 titled *Monografia Remanufacturing and Advanced Machining Processes for New Materials and Components* in amount 70,073.75 zł.

Library of Congress Cataloging-in-Publication Data

Names: Gevorkyan, E. S., author.
Title: Remanufacturing and advanced machining processes for new materials and components : remanufacturing and advanced machining processes / E.S. Gevorkyan, M. Rucki, V. P. Nerubatskyi, W. Żurowski, Z. Siemiątkowski, D. Morozow, A. Kharatyan.
Description: First edition. | Boca Raton, FL : Taylor & Francis, 2022. | Includes bibliographical references and index. | Summary: "Remanufacturing and Advanced Machining Processes for Materials and Components presents current and emerging techniques for machining of new materials and restoration of components. It also examines contemporary machining processes for new materials, methods of protection and restoration of components, and smart machining processes. It presents innovative methods for protection and restoration of components primarily from the perspective of remanufacturing and protective surface engineering. The book is aimed at graduate-level students, researchers, and engineers in mechanical, materials, and manufacturing engineering"-- Provided by publisher.
Identifiers: LCCN 2021037850 (print) | LCCN 2021037851 (ebook) | ISBN 9781032111568 (hbk) | ISBN 9781032111575 (pbk) | ISBN 9781003218654 (ebk)
Subjects: LCSH: Machining.
Classification: LCC TJ1185 .G477 2022 (print) | LCC TJ1185 (ebook) | DDC 621.9/02--dc23/eng/20211029
LC record available at https://lccn.loc.gov/2021037850
LC ebook record available at https://lccn.loc.gov/2021037851

ISBN: 978-1-032-11156-8 (hbk)
ISBN: 978-1-032-11157-5 (pbk)
ISBN: 978-1-003-21865-4 (ebk)

DOI: 10.1201/9781003218654

Typeset in Times
by Deanta Global Publishing Services, Chennai, India

Contents

Authors vii
Abbreviations xi
Nomenclature xv
Introduction xvii

Chapter 1 Contemporary Machining Processes for New Materials 1

1.1 New Materials and Approaches to the Product Life Cycle, Remanufacturing, and Reuse 1
1.2 Foam Injection Molding 6
1.2.1 Technical Advantages of Foam Molding 7
1.2.2 Polymer Foaming and Processing 8
1.2.3 Foam Injection Molding Equipment 13
1.3 Gasodynamic Densification of Casting Materials 14
1.4 Diamond and Electrical Discharge Diamond Grinding 21
1.5 Electrical Discharge Machining Methods 24
1.6 Electrochemical Machining Methods 29
1.7 Ultrasonic Machining Methods 32
1.8 Electron Beam Machining 37
1.9 Ion Beam Machining 41
1.10 Laser Material Processing 43
1.10.1 Classification of Lasers 44
1.10.2 Laser Beam Machining Methods 49
1.10.3 Laser Surface Treatment 53
1.10.4 Remanufacturing with Laser Cladding Technology 55
1.10.5 Future Development of Laser Applications 57
1.11 Processing with Plasma 57
1.12 Waterjet Machining 65
1.12.1 Waterjet Machining Principle and Influencing Factors 65
1.12.2 Main Components of a Waterjet Machining System 70
1.12.3 Main Advantages of AWJ Machining 71
1.12.4 Limitations and Environmental Impact of AWJ Machining 72
1.13 Trends in Joining Technologies 73
1.13.1 Remarks on General Trends 74
1.13.2 Promising Joining Technologies 75
1.14 Methods Increasing Machining Accuracy 78

Chapter 2 Contemporary Methods of Protection and Restoration of Components 83

2.1 Application of Additive Manufacturing Methods along the Product Life Cycle 85
2.1.1 Short Review of AM and Hybrid Methods 85
2.1.2 Materials Processed by AM 92
2.1.3 AM Methods Related to the Product Life Cycle 96
2.2 Protective and Technological Coatings for Hot Working Processes and Heat Treatment 101
2.2.1 General Requirements 101
2.2.2 Starting Materials for Coatings 102
2.2.3 Composition of Coatings 106
2.2.4 Temporary Protective Coatings in Micromachining 108
2.3 High Temperature Coatings 108
2.4 Thermal Spray Coatings 112
2.5 Vacuum Plasma Spraying 118
2.6 Processes of Hard Alloys Coatings 120
2.7 Titanium Nitride Nanoceramic Matrix Composite Coatings 124

Chapter 3 Smart Machining Processes 127

3.1 Computer-Integrated Manufacturing 127
3.2 Smart Manufacturing 131
3.3 "Smart" Tools and Materials 135
3.4 Superplasticity and Its Application to Metal Forming 140
3.5 Severe Plastic Deformation in Nanostructural Materials Processing 145
3.6 Product Life Cycle Models and Potential Cultural or Social Implications of Repair 151
3.7 Concluding Remarks 154

References 155

Index 183

Authors

E. S. Gevorkyan

Doctor of technical sciences, Professor of Department "Quality, standardization, certification and manufacturing technology of products" at Ukrainian State University of Railway Transport, Kharkiv, Ukraine. E. S. Gevorkyan graduated with award from K. Marx Yerevan Polytechnic Institute in 1982 and after several years of industrial experience became a postgraduate student and junior researcher at Institute for Superhard Materials of National Academy of Science of Ukraine, Kiev, where he prepared his thesis in 1990. From 1992 to 2003 he was a senior researcher and associate professor of National Technical University "Kharkov Polytechnic Institute" (Ukraine), and since 2003 he is with Ukrainian State University of Railway Transport, Kharkiv, Ukraine. He was awarded the scientific degree in 2008, and was assigned an academic status in 2010. Research interests include the development of composite materials based on multifunctional purpose powders of refractory materials and metals and manufacturing methods based on sintering.

By the end of 2020, the total number of publications authored or co-authored by E. S. Gevorkyan was 286, including 240 articles, 46 conference abstracts, 32 patents (including US and EU patents), and 5 textbooks published in Ukrainian, approved by the Ministry of Education of Ukraine.

In 1994–1996 he took part in the project INTAS "Research of Properties of New Composite Cutting Tool Materials." In 2000–2003 he was a lead for the project STCU 2605 (NASA) "Development of Composite Material to Clean Emissions," financed by the United States and Canada. In 2019 E. S. Gevorkyan was a visiting professor at Yibin University – Vocational and Technical College (China, Sichuan, Ibin), and from 2019 till present a visiting professor at Kazimierz Pulaski University of Technology and Humanities in Radom (Poland, EU).

M. Rucki

After graduating in mechanical engineering, M. Rucki received a Polish Government Scholarship for PhD Studies, Poznan University of Technology (Poland), from 1994 to 1996. He obtained a PhD in metrology and measurement systems in 1997. He was an assistant professor at Poznan University of Technology, Institute of the Mechanical Engineering, Division of the Metrology and Measurement Systems, from 1997 to 2010. At present, he is working as a professor at Kazimierz Pulaski University of Technology and Humanities in Radom (Poland). In 2017–2019, he was awarded by the rector for his research activity and achievements.

V. P. Nerubatskyi

Graduated from the Ukrainian State Academy of Railway Transport with a degree in "Electric Transport" and "Quality, Standardization and Certification" in 2009. In 2017, he obtained his PhD degree in engineering sciences and since 2018 has been working as an assistant professor for the Ukrainian State University of

Railway Transport, Kharkiv, Ukraine. In 2019, he became associate professor of the Department of Electroenergy, Electrical Engineering and Electromechanics. Areas of his research interests cover semiconductor power converters of electrical energy for electric transport and power supply systems, as well as the industrial composite materials from powders of refractory compounds and metals and generally powder metallurgy.

W. Żurowski

W. Żurowski obtained his PhD degree in mechanical engineering in 1996 at Kielce University of Technology (Kielce, Poland) and habilitation in 2012 at Kazimierz Pulaski University of Technology and Humanities in Radom (UTH), Poland. Since 2012, he has been working as a professor for UTH Radom, and in 2013–2016 he was a director of the Institute of Machine Building of the Faculty of Mechanical Engineering, UTH Radom. At present he is a vice-rector for Scientific Researches at UTH Radom. He is a member of the Polish Tribological Society, South African Institute of Tribology (RSA), Society of Tribologists and Lubrication Engineers (USA), and several other scientific societies in Poland. W. Żurowski is the author or co-author of 14 books and handbooks, more than 80 research articles, and 40 conference papers.

Z. Siemiątkowski

Z. Siemiątkowski received his PhD in mechanical engineering and technology in 2000. In 2000, he became an assistant at Kazimierz Pulaski University of Technology and Humanities (UTH) in Radom (Poland), Faculty of Mechanical Engineering. He was the vice-dean of this Faculty from 2012 to 2019 and has won several individual awards for organizational and scientific achievements. Since 2019, he has been the head of the Faculty of Mechanical Engineering, UTH, Radom. Z. Siemiątkowski has been a coordinator of nine research projects conducted in cooperation with industrial enterprises and a leader of research tasks in further three projects financed by Polish governmental and ministerial institutions.

D. Morozow

D. Morozow received his PhD in mechanical engineering and technology in 2000. In 2000, he started research and didactical work at the Faculty of Mechanical Engineering, Kazimierz Pulaski University of Technology and Humanities (UTH) in Radom (Poland). He is a certified tutor on the Heidenhein and Siemens CNC centers, as well as MasterCAM technological design, and co-authored the relevant textbook. He participated in the organization of the scientific and didactic CNC laboratory at UTH. As a researcher, he took part in several research and technology projects with industrial enterprises, such as Pratt and Whitney, Celsa Huta Ostrowiec, and others. Since 2016 he has been a leader of the project "Modification of Ceramic Tool Materials with Ion Implantation" conducted at the Faculty of Mechanical Engineering, UTH Radom.

A. G. Kharatyan

In 1980, A. Kharatyan graduated with a diploma of excellence from the Faculty of Mechanics and Mechanical Engineering of Yerevan Polytechnic Institute (YPI), Armenia. Then he studied a postgraduate course and worked at YPI as a junior researcher and assistant professor. In 1985 he defended his PhD thesis in the field of mechanisms, machines theory, and robotics, and in 1986 he was appointed an associate professor in the chair of mechanics. Since 1992 he has been working for State Engineering University of Armenia (SEUA), Vanadzor branch, as the head of the mechanics chair, Deputy Director of Scientific Affairs in SEUA, Head of the Department of Basic, General Engineering Education of SEUA, and since 2009 until present he has been the Director of SEUA (NPUA), Vanadzor branch, Armenia. During the period from 1996 to 2015, he had a scientific and educational experience at the Brunel University London and Queen's University Belfast (UK), at the Technical University of Athens and Higher School of Engineering in Marseilles (France), at Lund University in Sweden and Royal University of Stockholm (Sweden), at the University of Koblenz-Landau (Germany), at the Link Campus University (Italy), at the University of Lublin (Poland), at University of Rotterdam and Utrecht (The Netherlands), and in the Austrian office of the World University Service, Graz (Austria).

In 1999 he founded the University-Industry Interaction Center of SEUA, Vanadzor branch, in a voluntary manner, and also the Students' Career Services and Production Relations office. He has received several governmental and ministerial awards in Armenia and in 2018 was elected a full member of the Engineering Academy of Armenia. He is an author of more than 50 scientific-methodological works, manuals, including 9 inventions, and more than 15 articles that have been published internationally.

Abbreviations

ABS	acrylonitrile butadiene styrene
ADCA	azodicarbonamide ($C_2H_4O_2N_4$)
AI	artificial intelligence
AM	additive manufacturing
AMT	advanced manufacturing technologies
APS	atmospheric plasma spraying
AR	augmented reality
CAC	carbon arc cutting
CCL	color center laser
CE	circular economy
CGC	composite galvanic coatings
CIM	computer integrated manufacturing
CM	chemical milling
CMC	ceramic matrix composite
CPS	cyber physical system
CVD	chemical vapor deposition
CW	continuous wave
DED	direct energy deposition
DIY	do-it-yourself
DMLS	direct metal laser sintering
DoE	design of experiments
DRAM	distributed recycling via additive manufacturing
DT	digital twin
EBM	electron beam machining
ECAP	equal channel angular pressing
ECH	electrochemical honing
ECM	electrochemical machining
ECMM	electrochemical micromachining
ECMP	electrochemical mechanical polishing
EDDG	electrical discharge diamond grinding
EDG	electrical discharge grinding
EDM	electrical discharge machining
EoL	end-of-life products
EPS	expanded polystyrene
EST	electric spark treatment
FDM	fused deposition method or fused deposition modeling
FEL	free-electron laser
FFF	fused filament fabrication
FGC	functionally graded coatings
FIB	focused ion beam
FIM	foam injection molding

FSAM friction stir based additive manufacturing
FSP friction stir processing
FSW friction stir welding
GCP gas counter pressure process
GTAW gas tungsten arc welding
HA hydroxyapatite
HALT high activity low temperature process
HAM hybrid additive manufacturing
HAZ heat affected zone
HCPS human–cyber–physical system
HFC high-frequency capacitive plasmatrons
HFI high-frequency inductive plasmatrons
HLTS heavy loaded tribo-systems
HPCS high-pressure cold spraying
HPT high-pressure torsion
HVAF high-velocity air-fuel spraying
HVOF high-velocity oxy-fuel
IBAD ion beam assisted deposition
IBD ion beam deposition
IBE ion beam etching
IBF ion beam figuring
IBM ion beam machining
ICP inductively coupled plasma devices
ICS industrial control systems
ICT information and communication technology
IM intelligent manufacturing
IMM injection molding machine
IoT Internet of Things
IT information technologies
JIT justin-time
LAHT low-activity high-temperature process
LCA life cycle assessment
LCG life cycle gap
LCI life cycle inventory
LCR laser cladding remanufacturing
LCVD laser chemical vapor deposition
LENS laser engineered net shaping
LFC lost foam casting
LM lean manufacturing
LOM laminated object manufacturing
LPBF laser powder bed fusion
LPCS low pressure cold spraying
LPVD laser physical vapor deposition
LSF laser solid forming
MAM metal additive manufacturing

MD-AM multi-dimensional additive manufacturing
MF-AM multi-function additive manufacturing
MG metallic glass
MIM microcellular injection molding
MMa-AM multi-material additive manufacturing
MMC metal matrix composite
MMo-AM multi-modulus additive manufacturing
MRR material removal rate
MSc-AM multi-scale additive manufacturing
MSy-AM multi-system additive manufacturing
MW microwave
MuCell microcellular foam technology
NC nanocrystalline
OT operation technologies
PA polyamide
PA-GF glass-filled polyamide
PAC plasma arc cutting
PACVD plasma-assisted chemical vapor deposition
PAW plasma arc welding
PBF powder bed fusion
PCM phase change materials
PEG poly(ethylene glycol)
PEO plasma electrolytic oxidation
PIII plasma immersion ion implantation
PLA polylactic acid
PV peak-to-valley sphericity error
PVD physical vapor deposition
RF radio frequency
RFID radio frequency identification
PMMA polymethyl-methacrylate
RSW resistance spot welding
RUM rotary ultrasonic machining
SCF supercritical fluid
SLM selective laser melting
SLS selective laser sintering
SM smart manufacturing
SMA shape memory alloys
SMIS smart manufacturing information system
SPD severe plastic deformation
SPF superplastic forming
SPS suspension plasma spraying
SS stainless steel
STL standard tessellation language
SZ stir zone
TBC thermal barrier coatings

TCLE thermal coefficient of linear expansion
TGO thermally grown oxide
TIA totally integrated automation
TMAZ thermomechanically affected zone
TS thermal spray
UAM ultrasonic additive manufacturing
VCIM virtual computer integrated manufacturing
UHTC ultra-high-temperature ceramic
UIT ultrasonic impact treatment
USM ultrasonic machining
USR ultrasonic surface rolling
USSP ultrasonic shot peening
UST ultrasonic surface treatment
USW ultra-short wave
UV ultraviolet
WAAM wire-arc additive manufacture
YSZ yttria stabilized zirconia
μPTA micro-plasma transferred arc
μUSM micro-ultrasonic machining

Nomenclature

A	amplitude of vibrations (mm)
C	material constant
E	Elastic modulus or Young's modulus (GPa)
E_{surf}	surface binding energy (eV)
F_0	ion dose (ion/cm^2)
f	frequency (Hz)
f_p	frequency of the pulses (Hz)
I	current (A)
k_φ	efficiency dependent on cutting angle φ
p	pressure (Pa)
P	power (W)
T_g	glass-transition temperature (°C)
t_p	duration of the pulses (s)
t_x	fragility criterion, relation of shear strength of a material σ to the applied normal stress τ
U	voltage (V)
v	velocity (m/s)
v_t	velocity of a tool vibration (m/s)
η	efficiency
σ	shear strength of a material (MPa)
σ_y	elastic limit (yield stress) (MPa)
τ	normal stress (MPa)
φ	cutting angle (°)

Introduction

Contemporary development of industrial technologies is possible not only due to new processes and materials with new functional and technical characteristics but also due to remanufacturing of end-of-life (EoL) products (Le et al., 2015) and surface engineering methods designed to protect components and to prevent surfaces from degradation (Prashar et al., 2020). These measures increase the lifetime of components and engineering systems and provide substantial savings of energy and resources. Further energy savings are achieved through intelligent manufacturing (Wang, Li et al., 2019) and optimal maintenance strategies focused on manufacturing sustainability (Huang et al., 2019). A trend can be observed toward coating technologies that restore the properties of used components and substantially improve the quality and performance of a remanufactured product (Zheng et al., 2019).

Manufacturing processes are directly concerned with the change of form, dimensions, and certain properties of a part being produced. A wide spectrum of processes can be applied to achieve the goal, such as cutting, grinding, bending, forging, heat treatment, gluing, brazing, welding, bluing, diffusion welding, industrial etching, electrolysis, deep and surface hardening, explosive welding, water-jet cutting, sandblasting, and treatment with high-frequency current. In the present book, the authors attempted to focus primarily on the restoration techniques of EoL products and some surface engineering processes that improve the wear resistance of the components. The reader will be able to familiarize himself with finding out relatively new manufacturing processes and recent modifications to the conventional manufacturing processes. Thus, in the first chapter, machining processes for new materials are described, such as lost foam casting, diamond and diamond spark grinding, ultrasonic machining methods, cutting and shaping of the materials with an ion beam, waterjet, laser, and plasma. Additional attention is paid to the recent trends in material joining techniques and the methods increasing the accuracy of machining.

In the second chapter, the contemporary additive and hybrid methods of protection and restoration of components are presented in the context of the product life cycle. Firstly, protective and technological coatings related to the initial stages of a product life cycle, such as hot forming processes and heat treatment, are described. Next, the methods aimed at prolonging working time such as thermal and vacuum spraying or beam coating methods are characterized and described. Special attention is paid to the hard alloy coatings and nanoceramic matrix composite coatings used for the components that work in harsh conditions. Among remanufacturing methods related to the EoL parts, the reader will find repairing and strengthening cladding techniques aimed at restoring the dimensions and strength of a worn component.

The third chapter is dedicated to the recent trends in industrial enterprises in the frames of computer-integrated manufacturing (CIM) and the increasing application of the so-called smart machining processes aided with artificial intelligence (AI). These techniques are supporting the intelligent utilization of manufacturing resources and adaptive adjustment of production processes that support the level of organization called smart factories. As indicated by Wu, Lu et al. (2019), "within the smart factory, digital world and physical world should be seamlessly integrated, realizing the decentralized management of resources, combining the intelligent data fusion method to process industrial big data." Flexible adjustment of the process can be realized, however, at the level of "smart equipment" or "smart tools." In this chapter, attention is paid to the recent application of the superplasticity phenomenon in metal processing, as well as some nanostructural materials processing technologies based on severe plastic deformation. In the concluding section of the third chapter, product life cycle models are described, and potential cultural or social implications of the circular economy concept are indicated.

The methods and techniques presented in this book provide some insight into the contemporary trends in industrial practice, emphasizing resources saving and possibilities of performance prolongation for components and engineering systems. Nowadays, companies are looking for sustainability strategies in order to meet the market demands and reduce their environmental impacts and resource-related costs (Blume et al., 2018). The efforts to reduce resource consumption and improve energy efficiency are motivated by health concerns and environmental responsibility (Liobikienė and Minelgaitė, 2021), since there are strong links between the environment and the economic and social benefits of resource-efficient and environment-friendly manufacturing systems (Liu et al., 2018). It has been found that machining offers great potential for the conservation of energy and resources and for the reuse of raw materials (Denkena et al., 2019). The methods and processes presented in this book, together with the description of trends and possible developments, can provide good support for the efforts of scientists and engineers.

1 Contemporary Machining Processes for New Materials

Some materials have been used by humankind for millennia, e.g., ceramics, glass, iron, and copper, while others emerged in the second half of the 20th century, such as polymers, engineering ceramics, and composites (Dobrzanski, 2013). The continuous process of search for and exploration of new materials and methods usually intensifies when the existing methods are not efficient (Kumar et al., 2018). Nowadays, global competition has resulted in the rapid development of new materials, processes, and products (Velayudham, 2007). In fact, the variety of materials has increased tremendously in the last decades; new and more severe service requirements and demand for lower costs have arisen as well (Schwartz, 2010). Some sources estimate that the number of available engineering materials is close to 100,000, and it is expected that new materials with certain properties will be produced on customer's demand (Dobrzanski, 2013). Thus, the term "new materials" is often associated with materials that possess significantly different properties than the "traditional" ones, e.g., reduced weight, improved strength, toughness, wear and corrosion resistance, and safety for our health. The new generation of materials with improved properties poses problems to machining and hence requires new manufacturing processes and new machining approaches (Sreejith and Ngoi, 2001). In this chapter, we describe several material processing methods, starting from the foam injection molding and concluding with accuracy issues. But first we should discuss the main concepts driving today's industry.

1.1 NEW MATERIALS AND APPROACHES TO THE PRODUCT LIFE CYCLE, REMANUFACTURING, AND REUSE

It is not easy to define the term "new materials". When consulting Science Direct in February 2021, 138,112 results for "new materials" were found, out of which more than 14,000 were publications between 2020 and 2021. A proposed categorization suggests that the development of new materials covers the following:

1. Nano and microfabrication methods
2. Materials and systems mimicking nature, such as biomimetic materials and smart materials and systems
3. Materials for information storage and transmission, usually referred to as functional materials

DOI: 10.1201/9781003218654-1

Schwartz (2010) suggested that the following materials can be considered new:

1. Carbon–carbon composites
2. Shape memory alloys
3. Nanostructured materials, especially nanotubes, because fundamental physical, chemical, and biological properties of materials are surprisingly altered as their constituent grains are reduced to a nanometer scale owing to their size, shape, surface chemistry, and topology
4. Functionally gradient materials
5. Materials for micro-electro-mechanical systems and fuel cells
6. Liquid crystal polymers/interpenetrating network for polymers/interpenetrating phase ceramics

Thus, the term "new materials" can be associated with any materials recently developed that possess *significantly different properties* than the "traditional" ones (Sreejith and Ngoi, 2001). In recent years, however, more and more research into synthesizing new materials has been devoted to new functionalities, such as properties regulation, self-healing, reprocessing, solid-state recycling, and controllable degradation (Zhang et al., 2018). This indicates an awareness of the entire life cycle of the product. It is emphasized that without a rethinking of material utilization in the linear economy, many elements vital for industry, such as gold, silver, indium, iridium, or tungsten, could be depleted within the next 5–50 years (Tolio et al., 2017). Thus, apart from the introduction of new materials, the concept of circular economy (CE) should be implemented, which can be presented in the form of a 6R practical framework, namely reduction, repair, reuse, recover, remanufacturing, and recycling (Ghisellini and Ulgiati, 2020). Recycling is among the lowest in the hierarchy of EoL recovery strategy as it consumes much energy in melting and reprocessing and leads to material downgrade in terms of quality and usability in their subsequent life cycles (Wahab et al., 2018).

The vision of the circular economy paradigm is focused on fundamental changes in the current linear economic approach "take–make–dispose," which generates massive waste flows. It can be understood as an industrial system, restorative and regenerative by the very intention and design, aimed at keeping products, components, and materials at their highest utility and value along their life cycle (Tolio et al., 2017). In other words, CE replaces the product "end-of-life" (EoL) concept with restoration in order to eliminate waste through the superior design of materials, products, and systems. Circular economy may represent a new sustainable growth path and a business opportunity for the worldwide manufacturing industry. A growing number of the EoL products and resulting waste have made environmental protection and resource conservation an arduous task for countries across the world, requiring a series of governmental laws and regulations concerning remanufacturing to help support the development of such industries (Cao et al., 2020).

A sustainable transition to circular economy is expected to bring benefits in environmental, economic, and social terms. In environmental terms, CE practices have the potential to bring 80–90% savings in raw materials and energy consumption compared to the production of the same goods in the traditional linear model. The strong contribution to CO_2 emission reductions will positively affect climate change. In economic terms, major benefits for manufacturers cover the reduction of material and energy costs and of EoL material disposal costs. As a result, a general product price reduction of around 25–30% is possible, which, in turn, can boost the availability of high-quality affordable products, ensuring the increase of business competitiveness in emerging markets. In social terms, such benefits are expected of circular economy businesses as new jobs emerging from boosting consumption of sustainable products driven by lower prices (Tolio et al., 2017). The ratiocinating trend indicates that in the nearest future, greener technology that reduces wastage and reproduction, saves energy and the environment, and is easy to adopt is going to dominate (Jhavar et al., 2016).

Having the above-mentioned issues in mind, worldwide political and research agendas have started promoting CE strategies. In 2005, the G7 Summit Declaration of June 2015 issued the "Alliance on Resource Efficiency" document, while the European Commission launched the strategic initiative "Closing the Loop – An EU Action Plan for the Circular Economy." Since 2006, circular economy initiatives have been promoted in the Chinese "Five Year Plan" of development, and similar initiatives have been launched in the United States, Japan, and Australia (Tolio et al., 2017). However, at least three perspectives should be considered, namely, government policies, enterprises, and public awareness. Policies and regulations related to remanufacturing are elaborated, then advanced remanufacturing technologies are implemented by leading remanufacturing enterprises and national industrial parks, but public awareness on remanufacturing and citizens' participation in the environmental activities, both of which are beneficial to policy formulation and implementation, are or no less importance (Cao et al., 2020).

The European Remanufacturing Network is a pan-European project aimed at understanding and promoting the 6R practices, especially remanufacturing in the EU. On its website, remanufacturing activities are explained in the following way (*Remanufacture*, 2021):

- To return a used product to at least its original performance with a warranty that is equivalent to or better than that of a newly manufactured product.
- Involves dismantling a product, restoring and replacing components, and testing individual parts and whole product to ensure that it is within its original design specifications.
- From a customer viewpoint, a remanufactured product can be considered the same as a new product.
- Subsequent warranty is generally at least equal to that of a new product.
- Performance after remanufacture is expected to be at least equal to original performance specifications.

Place of remanufacturing in the life cycle of a product is shown in Figure 1.1.

Tolio et al. (2017), emphasize that sustainable transition to circular economy businesses will need to be supported with fundamental innovations in the manufacturing

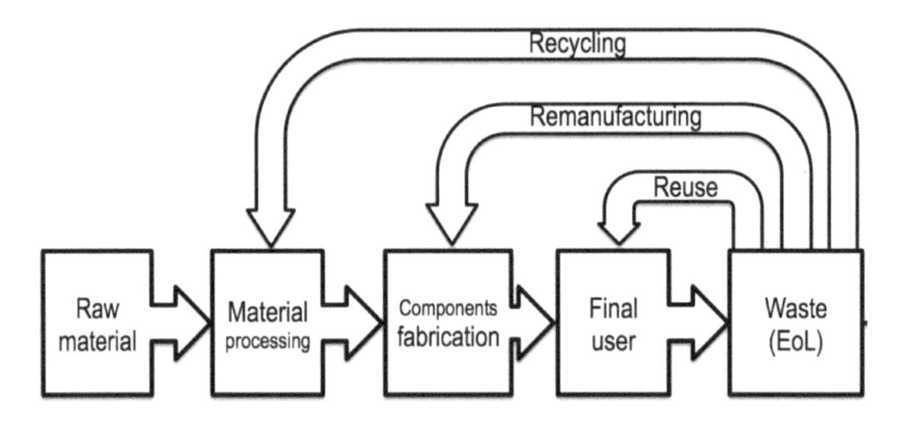

FIGURE 1.1 Place of remanufacturing in the product life cycle.

industry, at a systemic level, encompassing product design, value-chain integration, business models, involving demanufacturing and remanufacturing technologies. *Demanufacturing and remanufacturing* are fundamental technical solutions for an efficient and systematic implementation of circular economy that include *a set of technologies and systems, tools and knowledge-based methods to systematically recover, reuse, and upgrade functions and materials from industrial waste and post-consumer products, to support a sustainable implementation of manufacturer-centric circular economy businesses.* The authors have derived the following definitions of the main terms related to the CE concept from literature:

Reuse: a generic term covering all operations where a return product is put back into service, essentially in the same form, with or without repair or remediation.

Repair: correction of specified faults in a product. Repair refers to actions performed in order to return a product or component to a functioning condition after a failure has been detected, either in service or after discard.

Remanufacturing for function restore: it returns a used product to at least its original performance with a warranty that is equivalent to or better than that of a newly manufactured product. A remanufactured product fulfils a function similar to the original part. It is remanufactured using a standardized industrial process, in line with technical specifications.

Remanufacturing for function upgrade: the process of providing new functionalities to products through remanufacturing. Remanufacturing with upgrade aims to extend products' value life enabling introduction of technological innovation into remanufactured products in order to satisfy evolving customers' preferences and, at the same time, preserving as much as possible physical resources employed in the process.

Closed-loop recycling: recycling of a material can be done indefinitely, without properties degradation (upcycling). In closed-loop recycling, inherent properties of a recycled material are not considerably different from those of a virgin material, thus substitution is possible.

Open-loop recycling: conversion of materials from one or more products into a new product, involving degradation in inherent material properties (downcycling). In open-loop recycling, inherent properties of a recycled material differ from those of the virgin material in such a way that it is only usable for other product applications, substituting other materials.

Apart from reuse, remanufacturing, and recycling, three more EoL strategies can be found in the literature (Fegade et al., 2015):

Reconditioning: it is a process of restoring parts or components to a functional state but not above original specifications by resurfacing, repainting, etc.

Refurbishment: part or components are cleaned and repaired for reuse or resell.

Component Cannibalization: limited number of components are extracted from product and used to repair or rebuild another unit of same product.

Apart from benefiting the environment and ecosystem, the *remanufacturing strategy* has positively affected socioeconomies of various countries (Wahab et al., 2018). Being one of the EoL recovery strategies, remanufacturing is widely implemented in the transportation industry, namely aircraft, automotive, rail, and marine sectors. To date, the world's largest sector that benefits from remanufacturing is the automobile parts industry. In the aerospace/aircraft industry, original equipment manufacturers are involved in the remanufacturing process to a very great degree. It is known that many assemblies and structures such as engines, avionics, landing gear, and cabin interiors undergo remanufacturing procedures at least once. It is reported that some high-performance products such as Caterpillar's heavy-duty engines can be remanufactured as many as six times and serve seven use cycles. In the marine industry, intensity of remanufacturing is low compared to other transportation sectors such as aerospace, automotive, and rail. For example, marine components such as engines, propeller shafts, compressors, and pumps have been successfully remanufactured in many countries, but remanufacturing of large structures such as hull and vessels has not been reported thus far. Today, with a large number of ships approaching EoL, the share of remanufacturing should be increasing due to its positive impact on the environment (Wahab et al., 2018).

Used-up EoL products or components received for restoration are of a significant uncertainty and variability when it comes to their quantity and quality. In order to determine key characteristics of remanufactured products, the following steps should be undertaken (Lahrour and Brissaud, 2018):

1. Determining steps of the remanufacturing process
2. Proposing a framework of additive remanufacturing
3. Identifying key steps and characteristics of products that might be remanufactured by a chosen process

Remanufacturability evaluation should consider also effects of fatigue failure on the model of the remanufacturing critical threshold. Such models were analyzed based on a qualitative relationship between the fatigue life and remanufacturing requirements and compared with existing remanufacturing evaluation systems as reference, proposing two new remanufacturability evaluation indexes (Wang, Yu et al., 2020).

As far as selection of technology is concerned, the fuzzy logic approach can be applied in order to minimize vagueness in decision-making, thereby making results more similar to experts' thinking. Kafuku et al. (2019) evaluated and ranked appropriately six cleaning technologies, using criteria of technology cost, operating cost, and disposal effect. The fuzzy approach revealed that technology performance is largely impacted by criteria far beyond the technology itself, including purchasing cost, disposal cost, operating cost, and other support functions. Despite the fact that decision-makers can select a technology appropriately, the application of the fuzzy logic tool helps to accommodate vagueness, ambiguity, and subjective views of experts.

Application of remanufacturing to industry has an immense impact on costs and environment which is possible to assess. For example, Xiao et al. (2018) used life cycle analysis to compare environmental impacts and cost of a newly manufactured loading machine (S1) with its counterparts remanufactured at the original factory (S2) and at regional dealers (S3). It was demonstrated that the financial cost in S2 and S3 scenarios was 48% and 35%, respectively, of the cost in S1, while the climate change impacts in a functional unit decreased by 72% and 80% in S2 and S3, respectively. Apparently, remanufacturing in both scenarios still has a better environmental and economic performance than manufacturing a new product.

In the following chapters, we will emphasize techniques and processes especially suitable for new materials and remanufacturing. Remanufacturing is totally different from the other EoL strategies as a product is restored to like new condition with reduced waste generation and less resource consumption. As a result, a significantly higher profit margin for producers is achieved due to reduced production cost and environmentally proactive benefits (Fegade et al., 2015).

1.2 FOAM INJECTION MOLDING

Introduction of pores into traditional materials is one of the novelties in material sciences and applications. As such, porous materials are a class of functional-structural materials with an optimal index of physical and mechanical properties and they include metals, ceramics, and polymers (Liu and Chen, 2014). Wood may be considered the oldest form of foam, since it is a naturally occurring foam of cellulose, but the first commercial foam was sponge rubber that was introduced in the 1910s (Ebnesajjad, 2003).

Technology of the foam molding has been known for more than 40 years, but it is continuously improved and modified. Some injection molding technologies facilitate more sustainable production, while others allow the production of components with new structures and properties, but foam injection molding offers the potential to combine both (Kastner et al., 2020). It is no wonder that the casting of various polymer materials with different sorts of structural foaming is attracting even more attention with time. Compared to bulky polymer materials, polymer foam exhibits low density, good heat insulation, good sound insulation effects, high specific strength, and resistance to corrosion (Jin et al., 2019). Foamed plastics may be flexible, semi-rigid, or rigid, and their densities may range from 1.6 to over 960 kg/m^3 (Rosato et al., 2004). In addition, they display easy recyclability, a better warpage behavior, higher stiffness-to-weight ratio, or higher energy consumption during biaxial bending tests (Kastner and Steinbichler, 2020).

The main advantage of foam casting of polymers consists in the unproportional dependence between bulk material density and the strength of final products. In particular, it has been demonstrated that structural foaming saves from 10% to 20% of an expensive polymer used to produce an element similar to a bulky one and showed ca. 70% higher strength (Dryakhlov, 2009). Other advantages of foamed plastics over non-foamed polymers include ease of foam formation, ease of molding, and lower dielectric constant (Ebnesajjad, 2003).

What is the essence of the foam casting process? The foaming process of polymers can be divided into three stages: cell formation, cell growth, and cell stabilization (Jin et al., 2019). First, a blowing agent is added to a molten polymer under certain conditions and dissolved in the polymer. Counterpressure in the cylinder prevents the rapid expansion of bubbles and foam formation. Next, the molten polymer is injected into a mold cavity, where it occupies ca. 80% or 90% of the forming volume.

Now the counterpressure is absent, thermodynamic instability is generated from the supersaturation of a blowing agent. The foam structure is then produced by expelling the dissolved blowing agent from the polymer/gas mixture (Nofar and Park, 2018). The foam increases in volume and occupies all the space inside the mold. When it comes in contact with the colder walls of the mold, the high-quality surface of the produced element is formed. During further cooling of the molten polymer with the saturated blowing agent, residual pressure in foam cells promotes homogeneity of the structure. It should be noted that these peculiarities of the process require much smaller injection pressure and holding pressure, since a molten body possesses its own inner pressure. As a result, in the practice of foamed polymer casting, forces in the mold are reduced by 25% and even 50% compared to the parallel quantities in bulky polymers.

This technique can turn a majority of plastics into foams (Rosato et al., 2004). On the one hand, there are obvious savings in raw materials and energy. On the other hand, there is room for efficiency improvement to injection molding machines (IMM). For instance, it is possible to apply larger molds than nominally dictated by the mold clamping force, or to increase the number of cavities in a mold. Structural foaming often allows for increasing by up to 10–20 times the forming surface per 1 ton of the clamping force. This way, the range of produced element dimensions can be widened for the same IMM models (Vovk et al., 2018). However, limiting dimensions of molding machines should be considered, too.

1.2.1 Technical Advantages of Foam Molding

Apart from the above-mentioned functional advantages of foam plastics, the application of blowing agent substantially overcomes the standard processing limitations. Its benefits can be listed as follows (Dryakhlov, 2009):

- Considerable savings of the polymer material
- Formation of larger and more complicated products and components without a need for further machining or assembling operations

- Reduction of shrinkage and tightening defects and thus improvement of product quality
- Reduction of final product mass while keeping its strength
- Reduced requirements for the nominal clapping force and power of the related mechanisms
- Cost saving on instrumentation

In addition, the following benefits of economical production should be mentioned (Kastner and Steinbichler, 2020):

- Dissolved gases can reduce the viscosity of polymer melts by more than 50%, leading to lower energy consumption during dosing, longer flow lengths, and reduced injection pressures.
- Lower pressures, in combination with the absence of a packing stage, can allow for smaller clamping units.

However, it must be considered that bubbles in the polymer melt reduce heat transfer prolonging the cooling stage. Moreover, some blowing agents may behave like oxidants, affecting the tooling.

Nowadays, molds and their working parts may be fabricated out of aluminum and its alloys, reducing expenses. The durability of aluminum molds is shorter than that of steel ones, but it is sufficient to run 50 and even 100 thousands of cycles. In the case of small lot production, e.g., for experimental purposes, tooling can be applied based on epoxy matrix composites with metal frames and strengthening inserts.

1.2.2 Polymer Foaming and Processing

There are two different methods of obtaining polymer foams from gas production during the foaming process, namely with physical blowing agents and with chemical blowing agents. Properties of 23 groups of blowing agents and their selection, quantity, and technology of processing for 44 polymers can be found in (Wypych, 2017). The author also evaluates the importance of foaming parameters, including the amount of blowing agent, clamping pressure, die and mold pressure and temperature, delay time, desorption time, gas content, gas flow rate, gas injection location, gas sorption and desorption rates, internal pressure after foaming, saturation pressure, saturation temperature, screw revolution speed, and surface tension.

Chemical blowing agents cause the formation of gas bubbles due to their decomposition to solid and gas fractions at high temperatures. The agent is mixed with the molten polymer either prior to or during plasticization. Proportions of the chemical blowing agents solved in the polymer may be from 0.25 up to 5% by volume, depending on chemical composition and expected foaming effects (Vovk et al., 2018). The most important chemical blowing agents are ammonium bicarbonate (decomposition temperature 60°C), sodium bicarbonate (decomposition temperature interval 100–140°C), and sodium borohydrate (decomposition temperature 300°C). In some cases, water can be also used as a blowing agent (Feldman, 2010).

Chemical blowing agents are further classified into inorganic and organic categories (Annicelli, 2020). The organic blowing agents have advantages such as good dispersibility, stable gas output, and uniform bubbles (Jin et al., 2019). Depending on the nature of the process, foaming agents can be endothermic and exothermic.

Exothermic blowing agents release energy during a reaction, which must be dissipated through a plasticization unit and tool. After activation, the reaction runs its course without more energy being added and continues until the blowing agent completes its reaction (Rohleder and Jakob, 2016). The process is irreversible; it is not possible to control or to stop it. The pressure in gas chambers may reach 1.2 and even 1.5 MPa. Moreover, the process may not be completed when the product is removed from the mold as it requires a long time before finishing operations. The most often used exothermic foaming agent is azodicarbonamide (ADCA), although it decomposes at a relatively high temperature of 230°C (Kmetty et al., 2018).

When heat must be continuously applied in order to initiate and propagate a reaction, substances are referred to as *endothermic blowing agents*. They usually dissociate water when reacting, which can lead to a hydrolytic degradation of polymer chains (Rohleder and Jakob, 2016). The most commonly used endothermic foaming agents are inorganic. They are normally a mixture of sodium bicarbonate and citric acid, which can produce finer cells and reduce density from 1.24 to 0.645 g/cm^3 in case of neat, extruded poly(lactic acid) (Kmetty et al., 2018). When sodium bicarbonate or ammonium bicarbonate are used as blowing agents, peak pressure in bubbles does not exceed 0.8–1.0 MPa. The heat is distributed steadily and consumed by the polymer structure, so that the gas release effectively stops as soon as the heating is reduced. In addition, decomposition products of the inorganic foaming agents are harmful to neither humans nor the ozone layer and may therefore be applied in the food industry, e.g., for packaging production.

Chemical foaming agents are generally low-molecular-weight compounds supplied as powder or pellet. Their advantage over physical (gaseous) foaming agents lies in that they can be added to a solid polymer before heating, while physical foaming agents must be injected into an already fluidized polymer (Niaounakis, 2015). Chemical foaming does not require design modification of the IMM; a standard machine can be used, with the only exception of reduced power requirements. The agents are mixed with the polymer at room temperature and atmospheric pressure. The chemical blowing agents are mainly thermostable substances that do not require additional equipment to be stored or transported.

However, chemical foaming does not provide for strict control of the foaming process parameters, so that cell dimensions may vary from 10 up to 200 μm. Another demerit is the ability of some chemical blowing agents to discharge active gases, e.g., ammonia, that may promote corrosion of low-alloy steels.

Addition of a chemical blowing agent to a polymer requires very precise dosing, which means a more expensive apparatus. Otherwise, variation of agent proportions in a polymer may cause the quality of manufactured parts to become unstable across lots.

Selection of a chemical foaming agent is mainly determined by a processed polymer type. Some agents may be applied quite universally, like ADCA, while others are more specialized and dedicated to particular groups of plastics.

Physical blowing agents in a liquid or gaseous state under pressure are injected into a molten polymer, mainly inside a plasticating cylinder. The counterpressure in the cylinder prevents the mixture from bubbling, so that gas is solved in the polymer molt reducing its viscosity. It should be noted, however, that the effect on viscosity is dependent on many factors. It was demonstrated that mass fractions of blowing agents propane and carbon dioxide below 2 wt% had little to no effect in regard to viscosity reduction of a polypropylene melt, but a mass fraction of 3.5 wt% resulted in significantly decreased viscosity values (Vincent et al., 2020). Injected into a mold cavity, a polymer melt is foaming due to the lack of counterpressure, forming pores in the polymer. Usually, supercritical carbon dioxide ($scCO_2$), nitrogen, and air, but also various ecological gas mixtures are applied as processing solvents. The Montreal Protocol and its subsequent amendments to the Vienna Convention for the Protection of the Ozone Layer have led to the introduction of "second-generation" blowing agents, such as hydrofluorocarbons (HFCs) or hydrocarbons. Zero ozone-depleting hydrocarbon blowing agents comprise *n*-pentane, isopentane, or cyclopentane (Höfer, 2012). Application of some other blowing agents, such as methylene chloride, becomes more and more limited under the air toxics legislation within various states considering it a suspected carcinogen (Kaufman and Overcash, 1993; Jimoda, 2011).

Among the main shortcomings of the physical foaming agents, the following should be named:

- The blowing agents, both liquids and gases, require specialized equipment to be delivered and stored. Moreover, gaseous agents must be turned to their liquid state before application in a plasticizer. As a result, manufacturing costs increase.
- Injection unit of the IMM should be redesigned so that injectors may work in it. That means additional investment into existing equipment in order to adapt it to foam injection molding.

Microcellular injection molding (MIM) is widely applied in many industries, due to its excellent flexibility and capability, good scalability and environmental benefits, low cost, and high efficiency (Zhao et al., 2020). It emerged recently because the traditional FIM techniques were able neither to assure steady distribution of cells in the volume nor to provide strict control of the resulting density of a foamed material. Moreover, some foaming agents appeared to be environmentally harmful, and specialized FIM equipment was not suitable for the processing of bulky polymers.

To analyze the superiority of microcellular porous structures, Zhu et al. (2020) performed numerical simulations of static mechanical properties of polymethyl methacrylate (PMMA) microcellular foams. The authors demonstrate that for foams with the same average cell size (7 ± 1 μm), the compressive strength increases by 144% (the void porosity is from 65% to 37%); for foams with the same void porosity ($64 \pm 1\%$), the compressive strength rises by 42% (the average cell size from 21 to 8 μm). They also compare the structures with spherical, ellipsoidal, and polyhedral cells and find that the ellipsoids have a superior compression performance, with the compressive strength higher by 8.2% (Zhu et al., 2020). New research into partially melted crystal structures and high melt strength polypropylene demonstrates it is possible to decrease the average cell size to approximately 2 μm and increase the cell density to 1.49×10^{11} cells/cm^3 (Fu et al., 2020). There are also reports of changes from microcellular to nanocellular bubbles in poly(lactic acid) foam (Ni et al., 2020), but many results for corresponding foam densities after cell size reduction remain either too high or close to the density of non-foamed materials (Okolieocha et al., 2015).

Microcellular Foam Technology MuCell® has been introduced by the company Trexel Inc. (USA) and is considered one of the most promising injection molding techniques (Szostak et al., 2018). Influence of the processing parameters on cell structure and mechanical properties has been investigated in a wide range of plastics, such as PP, PC, PS, PA6, PLA, LDPE, PET, and ABS (acrilonitrile-butadiene-styrene), as well as reinforced polymers with nanoclay, talc particles, or glass fibers (Gómez-Monterde et al., 2015). It can be stated that the introduction of microcellular foam technology has initiated an important new development stage in the processing of most polymer materials.

In the microcellular technology, the application of a supercritical fluid (SCF) is crucial. SCF is defined as a substance for which both pressure and temperature are above critical and a chosen SCF must be non-toxic, non-flammable, and chemically inert, and its residues should be easily removed (Villamil Jiménez et al., 2020). Commonly available atmospheric gases, such as carbon dioxide (CO_2) or nitrogen (N_2), are used as blowing agents. Foaming with SFCs is conceptually simple, but in practice it is a complex dynamic process requiring a full appreciation of the thermodynamics, physics, and chemistry of solutions and interacting species, polymer sciences, and process engineering (Di Maio and Kiran, 2018). A supercritical fluid is injected with high precision into a molten plastic in the plasticating cylinder of the injection molding machine. The gas and the plastic are mixed together at continuously controlled SCF injection pressure, polymer molt pressure in the SCF injection area, and temperature of the molt in the plasticating section of the screw pump. It should be noted that both the plasticating section and the screw pump must be designed in a specific way different from that for bulky plastics, which has an effect on the overall costs of the method. The injected gas plasticizes the polymer upon its solubilization and reduces its apparent glass transition temperature or melting point to the processing temperature (Nalawade et al., 2006). It is reported that the addition of $scCO_2$ reduces the viscosity of PMMA by 70%, that of polystyrene by up to 60% at a processing temperature of 170°C, while the viscosity of polydimethylsiloxane (PDMS) is reduced by 60% at 50°C (Kazarian, 2000). Among others, this phenomenon allows for better filling of thin wall cavities, especially in case of flexible element formation.

This single-phase polymer-gas solution is quickly injected into a mold changing its thermodynamic state, so that gas bubbles separation begins. These form microcells with an inner pressure, required to fill in the mold cavity completely. As a result, holding pressure is almost not required, because in all “centers” of cells growth, pressure increases equally, even in the cavity areas distant from the IMM nozzle. Moreover, peak pressure in the mold can be reduced by 80% compared to the one prevailing in traditional injection molding. Due to the accelerated heat transfer from the polymer molt to the mold, the time of the cooling cycle can be reduced, which leads to the formation of equal-sized pores, steadily distributed in the molded part volume. Microcellular polymeric foams may have cell diameters ranging from 0.1 to 100 μm and cell density greater than 10^8 cells/cm^3 (Zhang et al., 2019). The cell size of most commercial microcellular plastics is in the range of 10–30 μm, and careful control of the process enables to produce foams with cell density above 10^9 cells/cm^3 and small and uniform cell sizes below 10 μm (Wong et al., 2016). In comparison, a typical conventional polystyrene foam will have an average cell size of about 250 μm, and a cell density of between 10^4 and 10^5 cells/cm^3 (Xu, 2010). For a wide range of polymers and fillers, the molded part thickness achievable in the MIM technology is reported between 1.5 and 4.0 mm (Sykutera et al., 2020).

Many advantages of microcellular plastics result from the foam injection molding technology. In particular:

- Polymer melt viscosity is reduced by up to 60%.
- Rib to nominal wall thickness ratio is increased.
- Reduced viscosity allows for processing at lower plasticization temperatures.
- Pressure in hydraulic units and injection pressure are reduced by 50%.
- The entire cycle time is reduced, especially due to short holding and cooling time.
- Mass of the product is reduced.
- Petroleum-based material consumption is reduced.
- Clamping force is reduced by 50–80%
- Sink marks, warpages, and areas of residual stress are prevented.

One of the most important benefits of the MuCell® technology is the fast production cycle. When minimal holding and cooling times are reached, the overall cycle can be cut by up to 40% (Trexel, 2021). Another important advantage is the material saving, since about 70% of the cost of a plastic part is the base material cost (Xu, 2010). Regarding MuCell, manufacturers state that weight reduction and corresponding material saving of up to 30% is possible, but for technically demanding parts it remains between 8% and 15% (Celanese, 2021).

Industrial practice indicates that adjustment of the IMM for a particular batch in the case of microcellular injection molding takes less time. Thus, batches with a stable quality can be produced in a shorter time, even in the case of thin-walled parts with a wall thickness of as little as 0.5 mm. It is noteworthy that conventional injection molding required some sort of compromise between cycle time and flatness of obtained surfaces free from sink marks, warpages, and concentrations of residual stresses. MIM technologies have made it possible to reach both these objectives for many product types, such as housings or fittings, without surface defects.

Among merits of the MuCell® technology, improved ductility, toughness, and impact resistance of the obtained microcellular material should be named, too (Wong, 2016). Most materials are considered suitable for microcellular foaming and competing companies develop new compositions according to market expectations. In addition, special modifications of MuCell® technology are proposed in order to optimize the process.

However, MuCell® technology has some limitations, such as reduction of mechanical strength or only matte surfaces obtained (Szostak et al., 2018). Some customers may not accept the roughness parameters or the microscopic surface discontinuities caused by bubbles. Nevertheless, the MuCell® process is primarily used in automotive, consumer electronics, medical devices, packaging, and consumer goods applications (Trexel, 2021), where roughness is desirable to prevent slippage. For this very reason, in some cases, sand blasting operations may be excluded from mold fabrication, providing further overall savings. When poor appearance is the issue, methods for surface quality improvement can be applied, such as the co-injection process, where a skin is injected over a microcellular part, or the heat and cool process, where control of mold temperature ensures continuity of the surface layer (Lima et al., 2016).

Another way to obtain an integral structure and an improved surface quality of a microcellular plastic is the gas counter pressure (GCP) process. The effect is achieved by pressurizing the mold to a pressure higher than that of the foaming gas prior to injection of the melt (Shotov et al., 1986). Prior to the injection, the mold cavity is filled with an inert gas at a certain pressure, and the polymer melt is injected against the gas pressure. Controlling the pressure inside the mold cavity, it is possible to prevent foaming during the skin formation process. When integral skin is formed, counterpressure of the inert gas is reduced starting the foaming process in the core of the produced part. Some reports say that under GCP control alone, when counterpressure was greater than 10 MPa, part surface roughness for transparent polystyrene (PS) was improved by 90% (Chen et al., 2013). The authors proposed a combination of the GCP and mold temperature control methods, producing molded MuCell parts with a high-quality surface, thin skin, defect-free surface, and small and uniform cell size, demonstrating the possibility of enhancing the application potential for the MuCell process.

It should be emphasized that the MuCell technology is not an alternative to the traditional injection molding, multicomponent sandwich-like molding for parts with continuous skin and foamed core, low-pressure foam injection molding, or molding with gas injection. Application of supercritical fluids can be treated as an additional technology which may be combined with others and promote further technological development. Nowadays, new injection molding foaming technologies are emerging, e.g., technology IQ Foam® (Gómez-Monterde et al., 2019), Optifoam®, Ergocell®, or ProFoam® (Xu, 2010).

1.2.3 Foam Injection Molding Equipment

FIM technologies in general, and MuCell in particular, are based on typical injection molding machine structures with modified aggregates and some additional equipment. A schematic of the MuCell system is presented in Figure 1.2.

In the microcellular IMM, the injection unit is principally different from conventional solutions. The plasticization unit must have a specific geometry to ensure

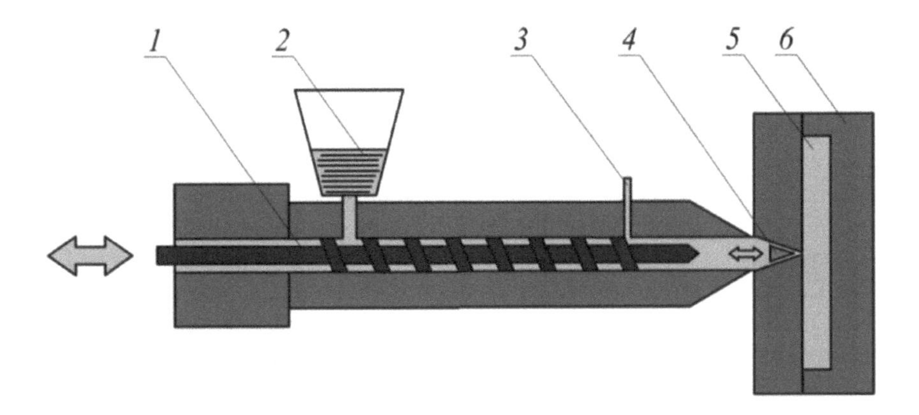

FIGURE 1.2 MuCell equipment: 1 – Reciprocating screw, 2 – Polymer feed, 3 – Inert gas injection port, 4 – Shut-off nozzle, 5 – Mold cavity filled with polymer-gas solution, 6 – Mold body.

a homogenous polymer–SCF solution and processing of traditional dense materials. The unit is equipped with special gas injection ports and additional pressure sensors to monitor polymer melt pressure in the cylinder. Other differences compared to conventional IMMs consist of an improved hydraulic system and control software. MuCell technology may use various drive units, e.g., hydraulic, electric, or mechanic. In any case, an additional module for SCF preparation, storage, and dosing is necessary.

At present, many international and local companies in North America, Europe, and Asia offer complete IMMs, based on their own technical solutions, able to perform high-quality microcellular injection molding processes.

1.3 GASODYNAMIC DENSIFICATION OF CASTING MATERIALS

Development of vacuum die casting technologies led to the expanded application of dry sand without binding substances. The sand molds are prepared by physical means, in particular, pressing the sand with a gas pressure drop. The pressure drop is able to transform loose sand into a stone-like substance around a mold cavity to be filled with molten metal. The process is sometimes referred to as the vacuum mold casting (*The Library of Manufacturing*, 2021), while the standard ISO 23472-1:2020 recommends the term "vacuum sealed molding" or the V-process.

In the thickness of the sand mold, vacuum is applied to the sand medium, while atmospheric pressure is applied to the top and bottom sand surfaces sealed with plastic films. Control factors of the V-process and molding sand, such as vibration frequency, vibrating time, degree of vacuum imposed, and pouring temperature may affect the quality of castings (Prasath Manikanda and Vignesh, 2017). In the modified process, where a pattern is made out of polystyrene foam, pressure from its gasification is combined with the atmospheric pressure. This is achieved in the lost foam casting (LFC) method or in combinations of LFC and low-pressure casting (Fan et al., 2014).

The main characteristic of the LFC process is that patterns made of polymers remain in the cast until the molten liquid metal is introduced. Contact with liquid metal causes intensive decomposition of the polymer pattern, which evaporates over a relatively short time simultaneously with metal crystallization (Prstić et al., 2014). During the process, the decomposed liquid products are pushed toward the molding cavity upper surface, then the liquid metal front. In order to achieve high-quality products, it is crucial to reach balance in the system of evaporable polymer pattern–liquid metal–refractory coating–sandy mold during the metal inflow, decomposition and evaporation of polymer pattern, and solidification of casting. Important factors influencing the process of decomposition and evaporation of the patterns in LFC are as follows: temperature, pattern density, type and thickness of the refractory coat layer the evaporable pattern is covered with, type and size of sand grain, respective permeability of sand, gating, and other structural details. The pattern density and permeability of the refractory coating and sandy cast determine polymer evaporation velocity. In order to achieve desired quality and predicted characteristics of a part, critical process parameters and the type of alloy for casting should be determined for each particular pattern material (Prstić et al., 2014). Among the advantages of LFC, the following can be listed (Fan et al., 2014):

1. Low surface roughness *Ra* 6.3 to 12.5 μm, and high dimensional precision from CT5 to CT7. The time of further mechanical processing can be decreased by 40% to 50% compared to parts obtained by the traditional sand casting method.
2. The structure design of LFC casting is flexible, with no drawing pattern inclination and a final casting can be obtained whole. As a consequence, the core can be omitted and the pore structure can also be directly manufactured, resulting in great cuts in time and cost of processing.
3. Loose-sand compact modeling is adopted in the LFC process, and the sand has no binder. Therefore, the production process of castings is simplified and the sand can be fully reused, saving production costs.
4. The LFC process has potential for cleaner production, since the polystyrene (EPS) does no harm to the environment at low temperatures. Compared with the traditional casting process, the harm from noise, CO, and silica dust is significantly decreased, resulting in an improved environment.

When the melt fills the mold cavity, complex physical and chemical phenomena take place, including heat transfer, filling flow, chemical reactions, cooling, and solidification. These influence each other, so that the liquid metal filling process is very complex and ultimately affects the casting quality. In recent years, researchers have carried out extensive simulation and experimental research into the liquid metal filling process flow and heat transfer of LFC (Xie et al., 2015). Some drawbacks of the LFC motivate further development of several novel LFC technologies, such as lost foam casting under vacuum and low pressure, vibration and pressure solidification conditions, expendable shell casting technology, and preparation technology of bimetallic castings based on the LFC process (Jiang and Fan, 2018).

Lack of binders in the sand and removal of foundry gases from the mold by vacuum substantially improve working conditions in the foundry. Moreover, after the introduction of the no-bond methods of vacuum molding, foundry does not affect the environment and leaves comfortable conditions for human activity. The sand may be reused with a small amount lost, ca. 5% per cycle. It is cleaned and cooled usually using pneumotransportation outside buildings, saving the area and improving the optimization of the plant area.

However, in practical solutions of vacuum seal molding, especially when thin-walled components of high geometrical complexity are manufactured in rather small batches, some problems may emerge. Technology itself allows for the application of cheaper equipment and usually one vibration unit is projected. Its technical parameters are calculated considering an "average" mold for typical patterns expected for manufacturing at that stage, but rapidly changing demand compels changes to manufacturing programs. As a result, the vibration unit, unable to adapt to changing patterns of different mass, shape, and complexity, may become a bottleneck in the entire V-process.

During the process, preforms are assembled into a cluster, the cluster is invested in dry sand in a molding box, and the sand is compacted by vibration. The conventional filling of the vacuum forming flask with sand may cause deformation or abrasive wear of the pattern affecting the quality of a final product. This is especially important in the case of brittle patterns or patterns covered with non-stick powders. In addition, sand movement is accompanied by dusting, which may cause serious health problems and other violations of sanitary standards. To avoid these complications, the cluster inside the molding box is protected with a movable screen to provide a barrier to the sand stream. The frame-like screen inside the molding box covers the cluster and is moved up as the sand fills the box. The sand is poured along the perimeter of the screen between its stacks and the walls of the box, and the screen is moved up after the box is filled with the sand. A schematic of this solution is shown in Figure 1.3.

The pattern cluster (3) is placed in the molding box (1) on a layer of sand and then covered with the screen (4). The sand from the hopper (7) is poured through the flexible sleeve (5) with a distribution unit (6). As the sand is evenly filled around the perimeter of the screen, the latter is lifted, causing the sand to pour down freely at the repose angle of 33–38° to the pattern cluster. When the pattern is covered with powder or viscous paint, the sand is pressed against it or sticks to it as appropriate.

When brittle patterns are processed, the screen shape may be closer to a cylinder with a vertical side surface. Inside the screen frames, facing sand or liquid may be placed and the screen may be lifted after the entire box is filled. The circulating sand mixture after the casting may be reused. When low-temperature facing mixtures are used, the screens may be fabricated out of a heat-insulating material. Ice patterns that require low-temperature mixtures can serve as an example as the development of new types of cryotechnology for foundry increases its environmental cleanliness, replacing traditional polymer pattern materials with frozen water (Doroshenko et al., 2012). Investment casting with ice patterns is similar to that with wax patterns, a major difference being that an interface agent needs to be coated around an ice pattern to protect it from damage during the process (Liu and Leu, 2006). After the box is filled with sand and the screen is removed, the vacuum is released and the molten

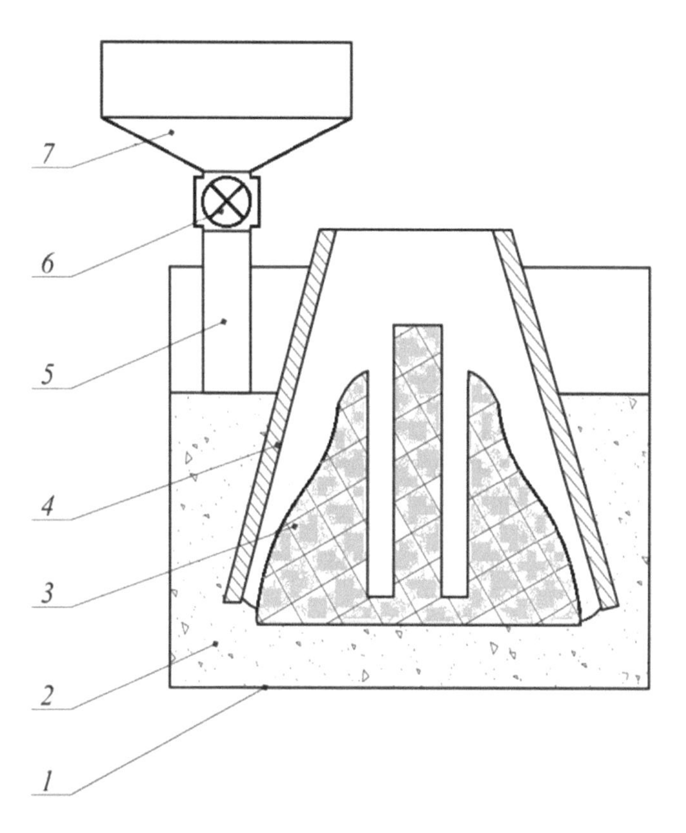

FIGURE 1.3 Protection of patterns from damage: 1 – Molding box, 2 – Sand, 3 – Pattern cluster, 4 – Protective screen, 5 – Flexible sleeve, 6 – Distribution unit, 7 – Hopper.

pattern removed leaving the mold cavity. While it is better to remove the conical screen (4) gradually as the sand pours in, the vertical-sided screen is easily removable from the full box when the sand is vibrated and friction between the sand and the screen is reduced. In any case, the movement of the screen should not exceed the critical velocity of the hopping motion of sand grains. If the sand grains' velocity remains below 0.039 m/s, then the fine fractions of 0.022 mm and greater do not contribute to the dusting.

The positive technical effect of this method is based on the prevention of the sand particles movement after their first contact with the pattern surface. As a result, casting quality is improved, especially in the case of brittle, thin-walled patterns, in particular ice ones. Limited screen velocity below its critical threshold prevents dusting of the fine sand particles, e.g., when a 1-m high sand layer is poured on the screen removal for 26 s and longer, fine particles cannot be put into hopping motion. It is impossible to keep sand particles below their critical velocity and thus prevent them from dusting when the sand is poured directly from the flexible sleeve.

Gravimetric results obtained for the molds from non-bonded sand under vacuum provide fundamentals for a novel gasodynamic densification method of casting materials (Doroshenko and Shinsky, 2009). It is demonstrated that when the air is blown through vibrated sand, its particles gradually move toward the vacuum filter. The smallest fractions move first, leaving the main sand body weakened or even destroyed. In the air promoted flow, the entire sand mass is moved until its particles meet the barrier. When the pressure drop is maintained, the air or liquid flow finds paths of the smallest resistance until these are clogged by air-carried fine particles. Additionally, when brittle, thin patterns are applied, the sand is not densified from the top, so that its upper layer is weaker than the lower ones. Moreover, the insufficient density of the sand without binders allows the molten metal to penetrate it, impairing the surface quality of the casting. There is also a problem from the perspective of energy savings, since the mass of vibrating components is close to the mass of sand and additional energy is consumed for the motion.

An interesting combined method of vibrating sand under vacuum is proposed by Doroshenko (2013). Among its advantages are simplified equipment, multifactor densification of sand, controlled gas pressure and vibration frequency, and wide possibilities for process automation. Figure 1.4 presents a schematic of the method.

The molding box (1) is filled with the sand (2) around an evaporable polystyrene foam or ice pattern (3). The upper surface of the sand is covered with a synthetic

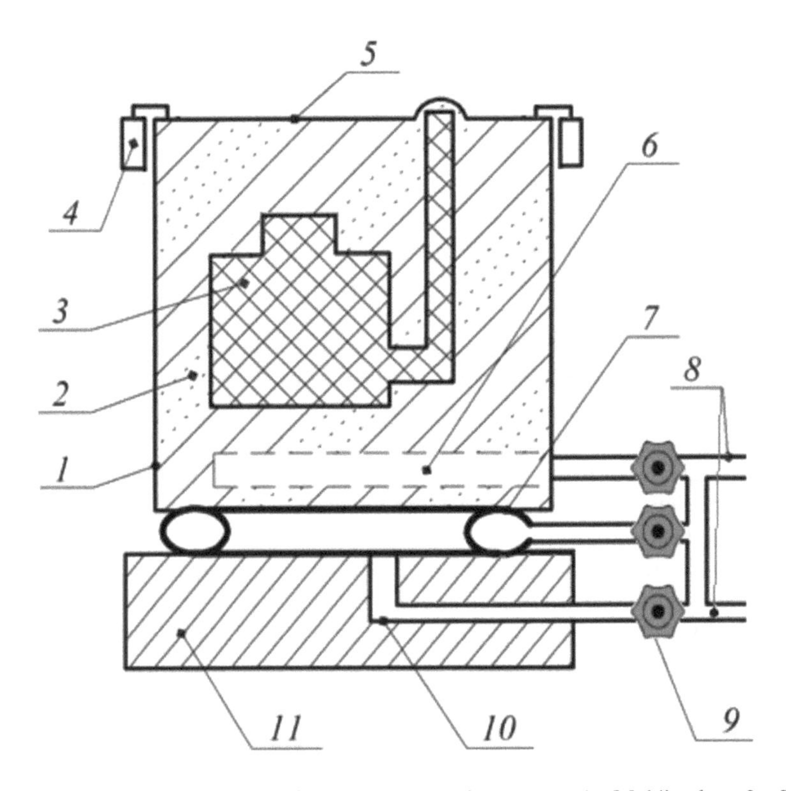

FIGURE 1.4 Combined method of sand pressing under vacuum: 1 – Molding box, 2 – Sand, 3 – Lost pattern, 4 – Clips, 5 – Synthetic film, 6 – Ventilation systems, 7 – Damper, 8 – Pressure distributor, 9 – Valve, 10 – Pipeline, 11 – Base.

film (5) fixed with clips (4). Vacuum is drawn through the ventilation system with a porous pipeline (6). The molding box is placed on the damper (7), a closed elastic pipe under pressure, made of 5 to 15 mm thick resin. The damper system is connected to the distributor (8), where the ventilation systems (6), pulse valves (9) and (11), controlled relays (1), and the pipeline (13) are connected as well. The latter delivers gas to an airtight cavity between the base (14), damper (7), and bottom of the molding box (1). The pipeline (13) and other parts are equipped with valves (12) or gas pressure reductors.

The sand is vibrated by alternating gas pressure in its porous mass. For that purpose, 2″ pulse valves can be applied, but for larger boxes it is better to use 2.5″ and 3″ valves. The valve (9) is connected with the vacuum pump that provides a pressure of 20 kPa and more. The valve (11), in turn, is connected to a pressured air source. When the film (5) is absent, the pressure is kept below the critical value to prevent the airflow with the velocity of hopping motion of sand grains, or the box is covered with a fine mesh to prevent dusting in the working area. Sand densification is performed through opening and closure of the valves (9) and (11) with the relay (1), which generates alternating pressure in the distributor (8).

Controlled frequency sinusoidal pressure in the sand generates important phenomena that contribute to high-quality compaction of the sand mold. First of all, the sand particles experience cyclical pressure that causes reactive forces of compression and expansion in the particle material as well as in the pattern. The pressure fluctuations affect surface phenomena and promote changes in wetting forces related to condensating moisture, inner friction, etc. Gas molecules move among the sand particles and put the finest ones into motion like at the initial stage of air-carried transportation. Excessive, above atmospheric pressure is limited to prevent gas flow through the open surface of the molding box, potentially able to generate hopping movement of the fine sand particles.

The sand is under gravitation force and under the pressing impact of alternating pressure. Mechanical vibration of the sand with inertia forces and reduced inner friction promote effective compaction of the sand mass and excellent reproduction of the pattern surface. When the conditions are maintained, enabling the proper effect of these factors on the sand mold, it is possible to simplify the equipment (Doroshenko, 2013). Simplified apparatus may be based on two valves only, connected with a relay, so that alternating pressure is generated both in the molding box and under its bottom placed on the damper. Connecting the valve (11) with the atmosphere, so that atmospheric pressure is introduced, is another possibility. Then the cycle consists of the pressure drop in the sand mass down to its minimal value, its rise up to maximum, and then decrease to the minimal value again, where the cycle started. Dampers (8) may be made of any sort of elastic material, but it is crucial to keep the entire surface integral or tighten it with additional material to ensure a hermetic cavity under the molding box. When the damper is filled with pressured air, as it is shown in Figure 1.4, it is possible to synchronize sinusoidal pressure changes in the damper, cavity under the box, and/or sand mass inside the box. The ventilation system (6) may be connected to the bottom of the box, then the pressure in the sand and under the box can be synchronized directly through a choke placed in the bottom of the box. In that case, there is no need for a third pipeline to be connected to the pressure distributor (3).

When above atmospheric pressure is applied to the sand, the film (5) may be blown up at the initial stage of the process. In the event, another protective film with small perforations displaced from the perforations in the first film is recommended. Then the excessive air may leave the box, while under vacuum the films are forced to the sand surface. After several cycles, however, the pressed sand prevents the air from blowing through and the films remain fastened to their surface permanently. This phenomenon helps visual assessment and regulation of the vibration frequency and other working parameters. After regulation, the film must be either motionlessly tightened to the sand surface or damped oscillations must be produced that finally lead to a tight adherence of the film to the sand surface. When the latter state is reached, it indicates the prevalence of pressing phenomena in the upper sand layers, where compaction of the sand prevents the air from penetrating it. Simultaneously, lower sand layers reduce gas permeability and redistribute the pulsation energy of the airflow to lower cavities, thus intensifying the shaking process.

On the other hand, the proposed technology may be further developed to achieve improved results and repeatability through full automation of the pressing process. In particular, Rusakov and Shinsky (2014) described the method of three-phase power activation to obtain a combined effect of vacuum, vibration, and gas pulses. The method divides each cycle into three stages of different gas states in the molding box. At the first stage, the gas pressure is changed to produce a vacuum of 40–10 kPa. At the second stage of each cycle, a control receiver of a certain volume and pressure is connected, so that equilibrium pressure is reached between the box and the receiver. The third stage provides the atmospheric pressure to the box. During the cycle, the pressure is measured both in the sand and in the receiver, and from the measured values the overall void volume in the sand is calculated. The data is analyzed to assess the effectiveness of compaction and to determine the finishing time of the process. The system is able to generate vertical vibrations of the molding box with frequencies from 5 to 100 Hz and vertical acceleration between 0.5 and 0.95 *g*. The control and diagnostic system includes a frequency generator, gasodynamic vacuum block and gas pulse generator, and an automatic system for adaptive control. Data collection and diagnostics are performed by sensors and executive modules connected to a computer system. When the program identifies unchanged void volumes in subsequent cycles, the system performs the last cycle to stop vibrations and eliminate resonant states.

An additional advantage of the gasodynamic processing of sand molds is its suitability for cooled or heated gas, or even gas with liquid components. Some examples can be found in the literature (Doroshenko and Berdyev, 2013). In the case of ice patterns, cooled nitrogen may be applied and the molding sand may be sprayed with liquid nitrogen. To heat the ice pattern before melting in the mold, a gaseous heat carrier can be added, or gaseous reagent in the case of CO_2 process. These gases may be supplied to the molding box at the final moment of the compaction process through the distributor (8) or the valve (11) as shown in Figure 1.4. However, if gas does not generate substantial expenses, it may be supplied during the entire operation through the valve (11).

As with all manufactured engineering products, castings suffer from errors of shape and size. In general, the so-called precision investment castings were found to be precise at small sizes, but rather poor at large sizes, while the pressure die casting and modern sand castings appeared to be highly accurate (Campbell, 2015). Castings are an integral part of many industries and involve more than 25% of the metal processes. In the 21st century, sand casting and gravity die casting are the two major industrial processes. However, an observation made from several foundries indicates that the V-process is implemented more frequently than investment casting. Several industries started adopting this process and found that it gives excellent yield and aesthetics with a near-net-shape finish. As a result of the process implementation, the machining cost is cut down drastically and the process proves to be environment friendly as well (Senthil et al., 2020). Thus, the vacuum-sealed molding seems to be one of the most promising processes with potential for further development.

1.4 DIAMOND AND ELECTRICAL DISCHARGE DIAMOND GRINDING

Nowadays, new constructional alloy steel with chemical additions enhancing its properties is widely applied, as well as tungsten, nickel, and aluminum alloys. These materials are attractive, but due to their increased tensile strength, low thermal conductivity, low plasticity, and other special properties, they pose difficulties during machining. Their carbon content increased from 0.1% to 0.7% abruptly reduces the machinability of steels with titanium, molybdenum, tungsten, vanadium, and chromium additions, so that the specific output of grinding falls by 30%. All types of carbides cause fast tear and wear of grinding wheel's abrasive grains. When steels with austenite structure are ground, their special output falls by 25% in comparison with martensite–troostite structures. Addition of silicon to alloys dramatically reduces grinding machinability, while alloying of metals with more than 2–3% of Mo and W also impairs it (Ivanova et al., 2018).

At the final stages of mechanical treatment of hard alloys and steels, organic-, metal-, and ceramic-bonded abrasive diamond machining is widely used (Smirnov et al., 2020). A typical bonded abrasive tool includes super hard grains (usually diamond or cubic boron nitride with monocrystalline grains or microcrystals of sizes ranging from one to several microns, characterized by a higher ductility and mechanical strength), fillers (usually corundum, silicon carbide, boron nitride, etc.), binders (sintered metal, electroplated, resin, ceramic, hybrid), modifiers (homogenizers, greases), and body (metal, ceramic, polymer) (Staniewicz-Brudnik et al., 2015). Porous metal-bonded diamond grinding wheels had an excellent grinding performance for hard-brittle materials (Su et al. 2006). Recently, the brazing superabrasive technique for diamond and cubic boron nitride has been considered as one of the most promising approaches to the fabrication of grinding wheels. Compared with conventional bonding methods, e.g., electroplating or sintering, a brazed tool has shown a high cutting performance, especially a reduced cutting force and lower temperature when grinding difficult-to-cut materials, such as carbides, optical glass, ceramics, aluminum alloys, and stone (Huang et al., 2021). There are also reports on the fabrication of a binderless diamond grinding wheel (BDGW) intended to prevent impurities from embedding during the machining of soft and brittle materials. The "binder" features, as well as the "grains," were produced on a single piece of chemical vapor deposition diamond, while the "grains" were ablated with a femtosecond laser (Qu et al., 2020).

Highly wear-resistant diamond grinding tools enable machining of virtually all structural materials with minimal expenses, high accuracy of shapes and sizes, and low roughness of machined surfaces (Ivanova et al., 2018). Diamond grinding and sharpening are particularly suitable for the restoration of worn cutting tools, like axial hard-alloy mills (Vasil'ev and Popov, 2015).

In the last two decades, diamond has been recognized as an alternative to various other available abrasive materials for grinding and finishing processes (Shekhar and Yadav, 2020). Thus, diamond grinding appears to be one of the most promising machining methods with good prospects. Nevertheless, grinding of cemented carbides based on titanium carbide (TiC) or titanium carbide–nitride (TiCN) with nickel and molybdenum matrix, with no additions of tungsten, is still very difficult even using diamond grinding wheels with Bakelite or metallic binders both in conventional and electrolytic conditions. The process is characterized by large cutting forces and specific work of grinding, increased specific output, and low efficiency. Interesting investigations were performed on the material removal mechanism of TiC/Ni cermet, including the diamond scratching test and the grinding-induced surface damage mechanics (Zhu et al., 2021). The results demonstrated that the material removal experienced plastic deformation, plowing, and fracture in the dynamic scratching process, where the material microstructure played a determining role in obtaining surface characteristics of the TiC/Ni cermet. To achieve a smooth surface and ductile material removal by the ultra-precision grinding process, dislodgement of hard TiC particles and surface relief formation were induced by varying the material removal rate between the binding phases and the TiC hard particle under the scratching of the diamond grits. This way, by varying machining parameters, nanometric surface characteristics were formed.

Most studies like the one by Zhu et al. (2021) are based on experimental research involving numerous variables with different methodologies, posing difficulties to comparative analysis of results. Theoretical models and simulations also provide limited insight into these phenomena and help to solve some particular technical problems. For instance, Habrat (2016) presented selected results of research in the field of grinding cemented carbide with the use of two different types of diamond grinding wheels during grinding of the ductile cemented carbide CTS20D workpiece material. The grinding speed, depth of cut, and feed rate were considered as input process parameters, and the ANOVA (analysis of variance) was employed to check the results for the developed model. It turned out the application of a resin bond grinding wheel provided significantly lower grinding force components during the process. On the other hand, Zhang and Xu (2019) proposed an integrated model based on the surface topography of grinding wheels and simulated the grinding process of cemented carbide. The simulation results were analyzed to obtain a surface roughness model and a specific grinding energy model based on an undeformed chip thickness distribution. Some rules for the influence of the grinding wheel surface conditions on maximum material removal rate were derived from the analysis, so that adjustments were introduced to improve the maximum material removal rate of the grinding wheel. The optimization results were verified through grinding tests of a cemented carbide. In turn, Cai, Yao et. al. (2020) proposed a model to predict the peripheral grinding force in the grinding process. The actual cutting depth, including the influence of the different grain paths of each element, was calculated to simulate the actual cross-section cutting area and the wear flat area. Through simulation of grinding force using the grinding wheel topography and the grinding kinematic directly, the influence of the workpiece residual was included. The authors claimed that the simulated grinding force was more accurate than the force predicted on the basis of the total material removal rate (Cai, Yao et. al. 2020). Even though these and similar results are helpful, it is impossible to abstract them into a general theory or model describing the effectiveness of the variety of grinding wheels combined with numerous hard alloys and cemented carbides to be machined. It seems, however, that among the most crucial issues related to the efficiency of diamond grinding are stable interaction conditions between the diamond grit and machined material, wear resistance of the grinding wheel, surface quality and integrity after machining, costs, and environmental impact.

Electrical discharge diamond grinding (EDDG), in popular use recently for processing electrically conductive materials, is an improvement (Rao et al., 2017). In EDDG, electrical discharge erosion similar to electrical discharge machining (presented in detail in Section 1.5) is combined with mechanical abrasion of diamond grinding, which improves the material removal rate and average surface roughness *Ra* significantly (Unune et al., 2018). In addition, erosion discharges during EDDG remove contaminations from the grinding wheel and sharpen the abrasive grains, prolonging the high-performance level of the tool. Balaji and Yadava (2013) indicated that the process prevents the generation of excessive temperature and thermal stress during the grinding. The principle of electrical discharge diamond grinding is explained in Figure 1.5.

EDDG is a complex machining process involving several disciplines of science and engineering to understand the random occurrence of spark and impact of the nonlinear behavior of a workpiece material due to temperature-dependent thermal properties (Balaji and Yadava, 2013). As a result, the diamond grinding process is improved in three respects: intensification of the machining, machinability improvement of difficult-to-cut materials, and automation of the entire process. In response to the environmental challenge, taking into account economic considerations as well, innovative anhydrous processes of diamond-spark grinding (DSG) of difficult-to-process materials are proposed, where the application of solid lubricants promotes savings of water resources (Gutsalenko and Rudnev, 2020).

The EDDG and DSG processes provide new perspectives on understanding relations and peculiarities of hard-alloy and cemented carbide grinding, microscale

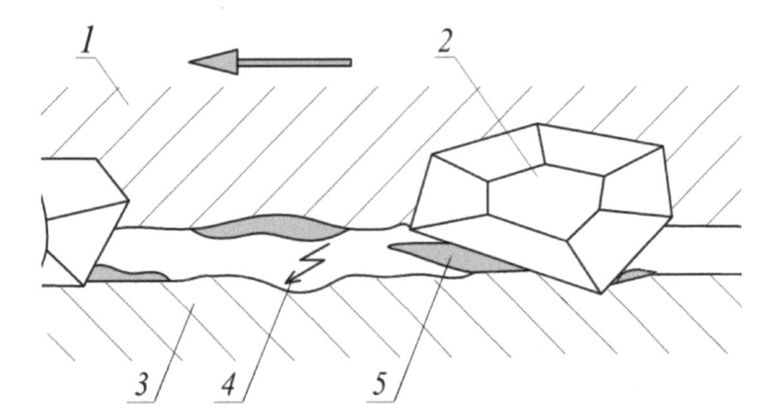

FIGURE 1.5 Schematic of the electrical discharge grinding process: 1 – Metallic bond of the grinding wheel, 2 – Diamond grit, 3 – Ground part, 4 – Electrical discharge, 5 – Chip.

physical and chemical interactions, and removal of infinitely thin layers of a material. The resulting models can describe efficiency and wear processes in order to improve the process, to stabilize working conditions and prolong the working time of diamond wheels, to normalize criteria of technical diagnostics and failure prediction, with a further adaptation of automated machining to the cyber-physical systems (CPS) or smart factory concepts.

Some theoretical and experimental results demonstrate that in certain conditions, a stable high material removal rate (MRR) is achievable, namely between 800 and 1,000 mm^3/min for tungsten-based cermets, and 600–800 mm^3/min for other cemented carbides. It is 4–5 times better than MRR for conventional grinding with metal-bonded wheels, and 2–2.5 times better than the electrical discharge grinding (EDG) process (Uzunyan, 2003). Similar improvement is demonstrated in other studies, where conventional electrical discharge machining (EDM) is compared to EDG with a rotating graphite wheel and then to EDDG with a copper bonded diamond wheel (Govindan and Praveen, 2014). In the case of machining of Al–SiC composite and titanium alloy, the MRR obtained by the EDDG process is about five times greater than for the EDM and about two times more than for the EDG process. The improvement is reached at low costs, high surface quality, and a high wear resistance of the diamond wheels. According to Uzunyan (2003), the cost of the EDDG process and the relative wear of diamond wheels can be reduced by between 25% and 50% compared to the conventional diamond grinding.

1.5 ELECTRICAL DISCHARGE MACHINING METHODS

As it was mentioned above, a method that utilizes electrical discharge in the form of arcs or sparks is a distinct machining process (EDM) with numerous modifications. Distinguishing between arcs and sparks is of no importance for electrical discharge machining (Schumacher et al., 2013). During the process, small amounts of electrical conductive materials are removed according to the thermal energy (melting and partial vaporization of a workpiece) produced by series of sparks occurring between the electrode and workpiece (Hourmand et al., 2017). The EDM process is potentially useful for machining materials with various hardness, complex shapes, strength, and temperature resistance.

EDM enables the machining of extremely hard materials, whereby complex shapes can be produced with high precision. Its inherent capability for automation is another feature fulfilling expectations of modern manufacturing. For these reasons, electrical discharge machining has become the most popular, nontraditional material removal process in today's manufacturing practice (Lauwers et al., 2012).

Two main EDM techniques may be distinguished, namely die-sinking EDM and wire cutting EDM. The die-sinking EDM uses an electrode to plunge into a workpiece and create a negative shape of the electrode, while the wire EDM utilizes a uniform wire that moves continuously and gradually cuts through a workpiece (Izwan et al., 2016). A schematic of the die-sinking EDM is shown in Figure 1.6. The tool electrode (1) and the workpiece (2) are immersed in a dielectric liquid and connected to the electric pulse generator (3).

Material removal is performed by applying direct current, pulsating (ON/OFF) with high frequency and forming square waves. The points of least resistance on the workpiece are attacked with numerous randomly ignited mono-discharges (7), each of them removes a small material amount (8). Extremely high temperatures from 8,000 to 12,000°C are created in the ionized column (Qudeiri et al., 2020), melting and evaporizing the material and leading to the formation of a discharge crater on the machined surface. A thin gap of about $\Delta = 0.025$ mm is maintained between the tool and the workpiece by a servo system (4). This distance is called the "discharge gap" and is usually kept between 0.005 and 0.1 mm (Qudeiri et al., 2020).

The amount of heat released on the electrodes is not equal and depends on polarity and pulse energy. Typically, the workpiece is the anode and the tool is the cathode

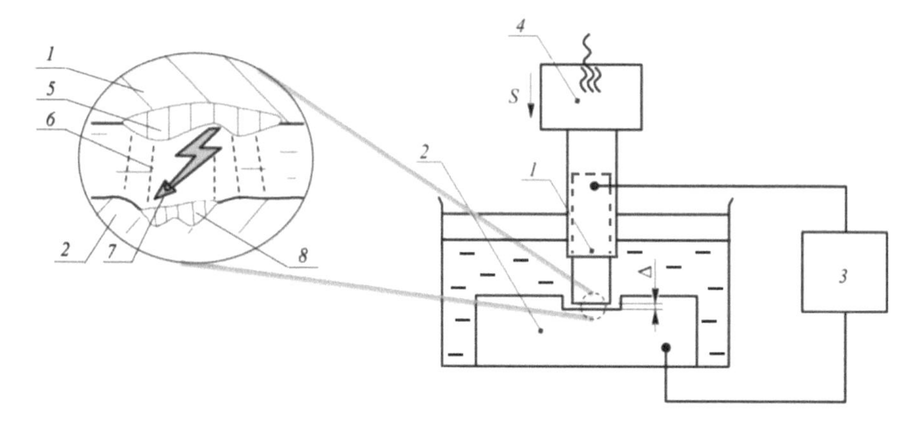

FIGURE 1.6 Schematic of the electrical discharge machining principle: 1 – Tool (cathode), 2 – Workpiece (anode), 3 – Power supply, 4 – Distance controlling device, 5 – Molten tool metal, 6 – Discharge column (ionized column), 7 – Electrical discharge, 8 – Molten workpiece metal.

(Palanikumar and Davim, 2013). It should be noted, however, that the workpiece (2) should be connected to the electrode where more heat is released, so that the material removal is to be maximized on the workpiece while the tool wear is minimal (Schulze, 2017). In particular, negative electrode polarity with a high MRR is recommended for high-precision machining, and in the wire EDM process, the electrode "wire" usually has a negative polarity to keep the machining rate high with a reduced tool wear rate (Qudeiri et al., 2020). Other major electrical parameters of the EDM process are as follows: discharge voltage, peak current, pulse duration and interval, and pulse wave form.

The entire process takes place in a liquid dielectric environment, such as kerosene, oil, or distilled water. Usually, two main types of fluids are distinguished, namely deionized water and the so-called dielectric oils based on hydrocarbon compounds (Uhlmann et al., 2010). The former is considered an ecofriendly substance since it does not release harmful gases, such as CO or CH_4, and yields a higher MRR and lower electrode wear (Rahman et al., 2014).

The dielectric fluid performs several main functions (Uhlmann et al., 2010):

- Isolation of the tool electrode from the workpiece electrode to achieve a high current density in the plasma channel
- Cooling down the heated surfaces of the electrodes and exerting a counter pressure against the expanding plasma channel
- Removal of particles after the discharge process preventing the particles from developing linkages between the electrodes that would cause process interruptions by short-circuiting or damage to the electrode surfaces

To perform these tasks, the dielectric liquid should display low viscosity, chemical neutrality toward tool and workpiece materials, intoxicity, safety, and low cost. Discharge erosion is determined by the chemical composition of the electrodes (tool and workpiece), dielectric liquid and energetic characteristics of electrical pulses, as well as by the process conditions. Thermophysical properties of a material, especially the melting and boiling temperatures, play a decisive role in the EDM and determine its machinability (Weingärtner et al., 2012). Generally, materials with lower melting temperatures can be eroded faster so that higher MRRs are achieved. If the machinability of a normalized steel is taken as a unit, then heat-resistant steels would have machinability of 1.3–1.4, while refractory metals and hard alloys would reach merely 0.4–0.5. Since hardened steels have thermal conductivity reduced, their machinability in the EDM process is 25–30% higher.

There are three main types of the EDM, depending on the electrode used (Palanikumar and Davim, 2013):

1. *Die-sinking EDM* consists of an electrode made according to a required shape and a workpiece submerged in an insulating liquid. During erosion, the shape of the electrode forms its inverse shape in the workpiece. This is an important accurate process normally used for making mold cavities. The machining speed depends upon the type of material, the area of the

material to be machined, and the machining conditions. Its principle is shown in Figure 1.6.

2. In the *wire EDM*, a wire electrode, usually 0.05 to 3.0 mm in diameter, is fed through an upper and lower diamond guide. Wire position is controlled by a computer program which contains the required path of the wire. The electrode is used only once and is discarded afterward. This form of EDM is the most accurate and can be used for both rough and finish machining (Palanikumar and Davim, 2013). A schematic of the wire EDM is presented in Figure 1.7.

A 0.05- to 0.3-mm wire (2) is a moving electrode without a specific shape. Its movement is usually performed along two coordinates, so that complex 2D contours can be produced with high accuracy, using short pulses of low energy. The wire is rewound continuously by the pulley and guides mechanism, while the workpiece (1) is moved in a perpendicular plane.

3. The third type of EDM is the *micro EDM mill*, where the electrode has diameters of 50 µm to 10 mm. The process is similar to the conventional die-sinking EDM, but the complex shape electrode is replaced by a set of standardized moving electrodes. The micro EDM provides for drilling pilot holes through heat-treated materials and carbides, e.g., deep cavities up to 10 mm with nearly vertical walls of 1.5° taper, or to machine very thin ribs below 0.1 mm. This technology is an important application in micromachining (Palanikumar and Davim, 2013).

From the perspective of distinctive kinematics, the *rotary tool EDM* should be mentioned (Dhakar and Dvivedi, 2017). In this process, presented in Figure 1.8,

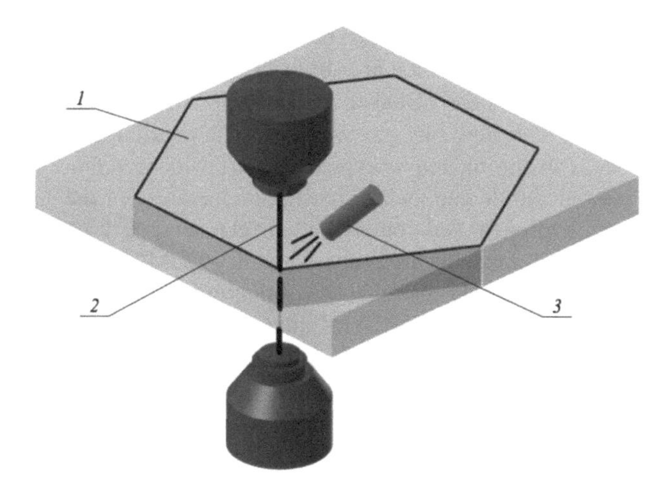

FIGURE 1.7 Schematic of wire EDM: 1 – Machined workpiece, 2 – Wire, 3 – Dielectric liquid supply.

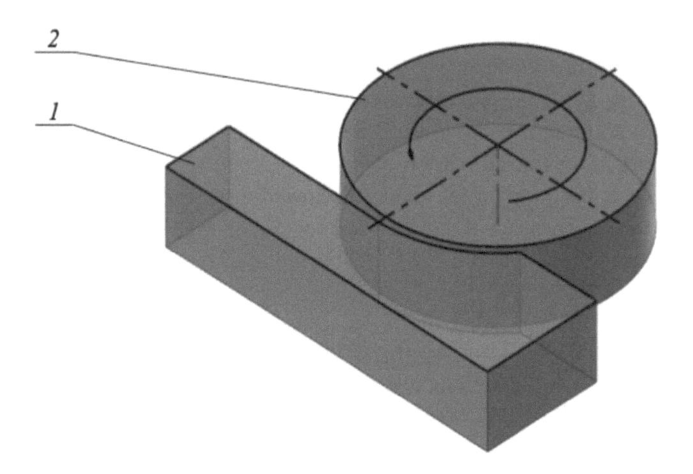

FIGURE 1.8 Schematic of rotary tool EDM: 1 – Workpiece, 2 – Tool.

combined movement of the workpiece (1) and profiled tool (2) is applied. This type is particularly useful for EDM dressing of shaped diamond wheels, machining of narrow grooves below 0.5 mm in steel and carbide rollers.

Besides, from the perspective of dielectric variations, powder-mixed EDM and dry EDM are distinguished as variants of the electrical discharge machining process (Qudeiri et al., 2020). A systematic review of the *powder-mixed EDM*, in which a fine abrasive electrically conductive powder is added to the dielectric, was published by Joshi and Joshi (2019). Suspended metallic powders in dielectric decrease its insulating strength and consequently improve the inter-electrode gap conditions, which enhances EDM performance and delivers a superior surface finish compared to the conventional EDM. The authors emphasized that lack of direct contact between the electrodes in EDM improves the process, because machining defects like mechanical stresses, clattering, and vibration are absent. *Dry and near-dry EDM processes* are sustainable variants of EDM (Dhakar and Dvivedi, 2017). The dry EDM process is conducted with a high flow rate gas as a dielectric medium supplied through a tubular electrode, causing molten workpiece material to be removed and flushed out from the inter-electrode gap and decreasing temperature of the sparking zone. Oxygen, nitrogen, hydrogen, and compressed air can serve as a gaseous dielectric fluid. Apart from environmental friendliness, the dry EDM is highly efficient with its MRR up to six times higher than that of conventional EDM with the same parametric conditions. In the near-dry EDM, liquid and air mist or a combination of gases like nitrogen, helium, argon, etc., are utilized as the dielectric medium. The two-phase dielectric medium provides stable machining, machining consistency, and better surface finish than the conventional EDM (Dhakar and Dvivedi, 2017).

Electrical discharge machining allows for accurate fabrication of complex and intricate profiles, holes with curved axes, thin-walled parts, narrow grooves and channels, and connecting holes in housings of hydraulic and pneumatic mechanisms. It is possible to produce thin, fragile sections easily and without burrs.

1.6 ELECTROCHEMICAL MACHINING METHODS

Electrochemical machining (ECM) can be traced back to the first electrolytic polishing process proposed by the Russian scientist E.I. Shpitalsky in 1911 (Koryagin et al., 2000) and developed by Gussev in 1929 (Ivanov et al., 2015). In the ECM process, controlled dissolution of an anodic workpiece takes place through electrolysis governed by Faraday's law, so that an approximately complementary image of the cathodic tool is reproduced on the anodic workpiece, dissolving it electrolytically without any deposition in the cathode (Jain and Pathak, 2017). Material is removed through anodic dissolution, an electrochemical process in which metal is dissolved from the anodic workpiece and gases are released on the cathodic tool as well as on the anode surface, and various electrochemical reactions take place in the bulk electrolyte (Bhattacharyya, 2015). In general, MRR depends on the electrochemical properties of the workpiece material, electrolyte properties, and characteristics of electric current. This can be seen from the equation (Ribeiro et al., 2013):

$$MRR = C \cdot I \cdot \eta \tag{1.6.1}$$

where C is a material constant, I is a supplied current, and η is a current efficiency. Values of MRR typically range from 1 to 20 cm^3/h depending on current density (Ribeiro et al., 2013). The ECM process generally uses low values of DC, between 8 and 30 V, while a high current density of 10–100 Acm^{-2} is generated, because the anode and a pre-shaped cathode are separated by a very small inter-electrode gap of 0.1–1.0 mm (Jain and Pathak, 2017).

Electrolytes are used to dissolve the anodic material and to flush away products of the electrochemical reaction. In addition, they remove heat from the passage of the current. There are four main groups of electrolytes (Leese and Ivanov, 2016):

1. Neutral aqueous salts
2. Aqueous acids
3. Aqueous bases or alkalis
4. Non-aqueous electrolytes

Aqueous salts are usually the first choice as they are generally inexpensive and tend not to cause damage to machinery setup. In contrast, an acidic electrolyte could corrode machinery over time. The most common electrolyte used for ECM is concentrated sodium chloride or sodium nitrate. For electrochemical micromachining (ECMM), a less concentrated electrolyte is required to enhance the machining precision by restricting the current passage through increased electrolyte resistance. Sodium chloride is regularly used to machine stainless steel when a bright surface finish is required, since it does not form a passive layer on the stainless steel surface. When close replication of the tool is more important than surface finishing of stainless steel, sodium nitrate is employed because it prevents stray corrosion, ensuring precise tool replication. However, when aqueous salt solutions do not provide an environment in which dissolution can occur, acidic or basic electrolytes can be used.

Acidic electrolytes are advantageous as the reaction products remain dissolved in the solution because the hydroxide ions produced at the cathode are neutralized by a high hydrogen ion (H^+) concentration. This allows to decrease the inter-electrode gap, as it does not get clogged with solid reaction products. Minimization of sludge also reduces the probability of sparks.

Alkaline electrolytes, such as sodium hydroxide (NaOH), are generally avoided as these can promote the formation of a passive film on the workpiece and, thus, require a larger inter-electrode gap. Potassium hydroxide (KOH) is preferable in some metal systems or in machining of tungsten carbide (WC).

Non-aqueous electrolytes are beneficial for passivating metals as they eliminate oxygen sources that form passive films. However, conductivities of nonaqueous electrolytes are low (Leese and Ivanov, 2016).

Flow rates of electrolytes are usually in the range of 10–50 liter per minute (lpm) (Jain and Pathak, 2017). A proper flow rate and circulation of an electrolyte in the gap regulate temperature and remove products of the electrochemical reaction (e.g., hydrogen and metallic ions), thus regulating the process. Joule heating increases MRR due to higher anodic dissolution kinetics and greater conductivity of the electrolyte. In turn, the release of hydrogen at the electrode decreases MRR because it dilutes the electrolyte concentration and diminishes its capacity to act as a current carrier. Metal ions from reaction products also hamper MRR if they are not properly washed away from the workpiece surface (Ribeiro et al., 2013).

Various kinematic solutions of the ECM find their areas of application (Koryagin et al., 2000):

1. Methods where electrodes do not move
   (a) Calibration
   (b) Contouring
   (c) Deburring
   (d) Edge rounding
   (e) Marking
2. With the translational motion of an electrode
   (a) Copying
   (b) Piercing of holes
   (c) Broaching
   (d) Calibration
   (e) Sharpening
3. With the rotational motion of a cathode
   (a) Processing of flat and shaped surfaces
   (b) Segmentation
   (c) Circular notching
4. With a rotating anode
   (a) Processing of shaped surfaces (external and internal)
   (b) Grooving (straight and spiral)
   (c) Segmentation

5. With a complex movement of the electrode
   (a) Cutting with wire (rod)
   (b) Cutting by means of the tubular-contour method

The ECM is perhaps most widely applied to copying of shaped surfaces and piercing of profiled holes, calibration of spline holes after heat treatment, and deburring with an accuracy of 0.2–0.3 mm (Koryagin et al., 2000).

The following are usually listed among the advantages of ECM (Saxena et al., 2018):

1. Independence of workpiece hardness.
2. Possibility to machine complex shapes.
3. Lack of tool wear and high surface finish as dissolution occurs at the atomic level.
4. Material removal rates can be controlled with electrical parameters (voltage, current, energy) and pulse characteristics (pulse frequency, time, duration, duty cycle).
5. Process forces and size effects don't affect results of machining.

At present, many variations of the ECM are implemented, but the oldest *electrochemical polishing process* (ECP) is still in use. The mechanism of the electrochemical polishing process is explained as follows (Łyczkowska-Widłak et al., 2020).

According to one hypothesis, the factor that has the greatest influence on surface smoothening during electropolishing is the emergence of a highly viscous layer on the anode as a result of the polarization of the processed material. The anodic diffusion layer is flat on the side facing the solution and the cathode, while on the side adjacent to the anode it takes the form of an anodic surface. This layer is characterized by a high electric resistance and is thinner on its micro-peaks than on micro-valleys. As a result, the peaks of roughness are dissolved first, because a higher density current passes through them. According to another hypothesis, acceptors of the metal dissolution process, such as particles of water and anions, play an important role. Water causes hydration of metal ions, detaching them from the surface, promoting the diffusion of the acceptors toward the surface of electrochemically polished elements. During electropolishing, the gradient of concentration of the acceptors increases at micro-peaks, so that they are the first to be subjected to anodic dissolution, while in micro-indentations, the dissolution process is slightly delayed until the metal ions close to micro-peaks are released from the metal surface (Łyczkowska-Widłak et al., 2020). Electrochemical polishing has been widely used for fine polishing of metals, with an initial average surface roughness *Ra* to the order of 1 μm down to a mirror finish (Chang et al., 2019). Some studies report that electropolishing results in nanometer size roughness (Allain and Echeverry-Rendón, 2018), while in the case of 2D-layered materials, the ECP provided results down to a monolayer (Sebastian et al., 2019).

In order to overcome some shortcomings and further improve the polishing process, *electrochemical mechanical polishing* (ECMP) technology is employed that combines electrochemical with mechanical action (Mohammad and Wang, 2016). As part of the ECMP, a passive film forms on the anode (workpiece) surface given certain electrochemical parameters due to anodic dissolution. This film is scratched at high spot regions of the surface with abrasives, then fresh layers of the substrate are immediately exposed to the electrolyte. Approximately 90% of the material is removed through electrochemical action while the mechanical abrasion is reduced to a minimum. The mechanical action increases the removal rate in the high spot region of the anode surface and accelerates the rate of anodic smoothing. Because only a small amount of material is removed mechanically, the abrasive tool life is ca. ten times longer than that of a conventional mechanical tool (Mohammad and Wang, 2016).

In terms of *remanufaturing technologies*, interesting reports are published on *electrochemical honing* (ECH) applications to improvement of surface quality and geometrical accuracy of recovered cylindrical shafts (Singh and Jain, 2016). In this process, most of the metal is removed on the atomic scale by anodic dissolution, while the honing process acts as a performance multiplier. Before ECH, the twin wire arc technique was applied to discarded surfaces of cylindrical shafts, recovering them with SS-316 material. The authors demonstrate that ECH of recovered surfaces provides a glazed texture and average surface roughness of 0.347 μm after the processing time of 90 s.

Machining and micromachining can be also performed with purely chemical methods, such as *chemical milling* (CM). CM is a subtractive machining process using baths of temperature-regulated etching chemicals to remove material for producing a required shape and blind features like pockets, channels, etc. (Bhattacharyya and Doloi, 2020). It also removes material from the entire surface of components for the purpose of weight reduction. This method is extensively used on metals with removal depths in excess of 10 mm (Tomlinson and Wichmann, 2014), although other materials are gradually becoming more essential.

The CM process consists of five steps: cleaning, masking, scribing, etching, and demasking. Solvent-based maskants can be applied by spraying, brushing, or direct immersion. Today, several new technologies have emerged regarding chemical milling maskants, so that recovery rates of 95% are available. Etchant aluminum regeneration performed today is 100% effective (Tomlinson and Wichmann, 2014). The efficiency of the chemical milling is rather low, between 0.4 and 1.2 mm/h, but in the case of large surfaces, CM appears to be very effective (Koryagin et al., 2000).

1.7 ULTRASONIC MACHINING METHODS

Another efficient and economical process for precision machining, especially of glass or ceramic materials, is ultrasonic machining (USM). Unlike other nonconventional processes, ultrasonic machining does not thermally damage the workpiece, which is important for brittle materials in service (Thoe et al., 1998). Compared to EDM and ECM, USM is advantageous because it is applicable to dielectric materials, too.

Cong and Pai (2013) emphasize the following advantages of this burrless and distortionless process:

- The process does not produce a heat-affected zone and residual stresses on machined parts and may also increase the high cycle fatigue strength of a machined part.
- Complex three-dimensional contours can be machined as quickly as simple ones.
- Dimensional accuracy of 5 μm and surface roughness of 0.1–0.0125 mm can be achieved.
- The equipment is safe and easy to operate.

USM is a mechanical material removal process. High-frequency electrical energy is converted into mechanical vibrations with a resonant frequency, usually via a piezoceramic or magnetostrictive transducer. An excited vibration is subsequently transmitted through a horn in order to focus energy and to amplify the vibration amplitude before it is delivered to the tool. The tool, appropriately shaped and located directly above the workpiece, starts vibrating along its longitudinal axis. A slurry comprising hard abrasive particles in water or oil is constantly supplied into the machining area. The vibrating tool causes abrasive particles dipped in the slurry to be hammered on the stationary workpiece performing material removal. Parameters of this process are usually as follows (Sawant, 2020):

- Amplitude of vibrations: 15 to 50 μm
- Frequency of vibrations is 19 to 25 kHz, or according to Koryagin et al. (2000), even up to 44 kHz
- Abrasive material is usually Al_2O_3, SiC, diamond or, according to Koryagin et al. (2000), boron carbide

Material removal mechanism in USM is mainly attributed to brittle fracture, initiation and propagation of cracks, and chipping of brittle material. During the USM process, a sequence of single-point indentations occurs. The crack formation as part of the single-point indentation process can be summarized in the following steps (Cong and Pei, 2013):

1. A single sharp indenter produces a plastic deformation zone.
2. At a certain threshold, a median crack suddenly develops.
3. With the increase of load, the median crack grows.
4. During unloading, the median crack begins to close and lateral cracks form.
5. After unloading, the lateral cracks propagate toward the workpiece surface and may consequently lead to chipping.

USM can be applied to the machining of workpieces harder than 40–60 HRC that cannot be machined with traditional methods. The process tolerance range is 7–25 μm and holes of small diameters of up to 76 μm can be obtained (Sawant, 2020).

Koryagin et al. (2000) distinguish four machining processes based on ultrasonic vibrations shown in Figure 1.9:

a – machining with a loose abrasive for removal of small burrs below 0.1 mm and grinding of small parts weighing ca. 10–20 g,

b – dimensional machining of the hard and brittle materials with the abrasive slurry,

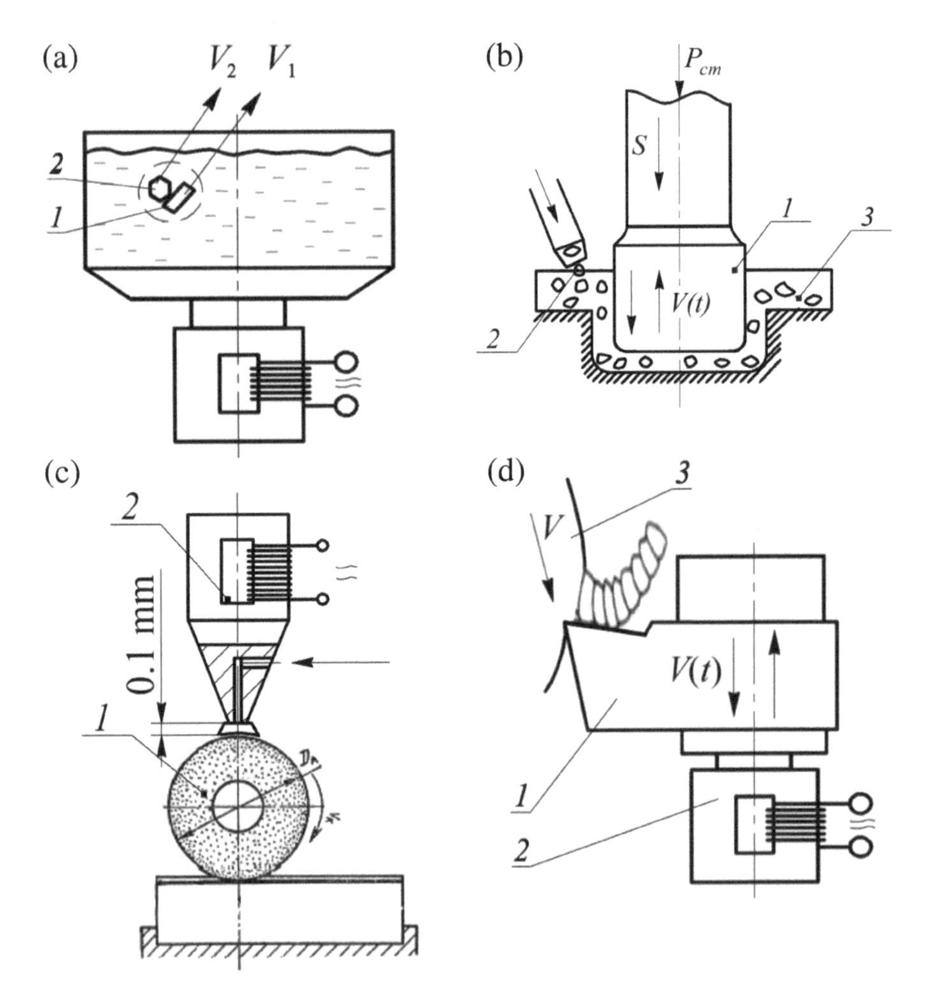

FIGURE 1.9 Ultrasonic vibration machining: *a* – Machining with loose abrasive (1 – Workpiece, 2 – Abrasive particle). *b* – Dimensional machining with abrasive slurry (1 – Tool, 2 – Abrasive, 3 – Liquid). *c* – Cleaning and lubrication of the abrasive wheel during grinding of viscous materials (1 – Abrasive wheel, 2 – Vibrator). *d* – Ultrasonic vibration-assisted cutting (1 – Cutting tool, 2 – Vibrator, 3 – Workpiece).

c – cleaning and lubricating of the abrasive wheel during grinding of viscous materials,

d – ultrasonic vibration-assisted (UV-A) conventional machining processes, where a small amplitude ultrasonic vibration is applied to either the cutting tool or the workpiece in order to intensify traditional cutting or grinding of difficult-to-machine materials.

In the first type of USM, as shown in Figure 1.9a, the workpiece is placed in the abrasive slurry subject to intense ultrasonic vibrations. Hydrodynamic flows cause the abrasive grains to have a velocity different from that of workpieces, due to the different densities. Thus, struck by abrasive grains, the workpiece material is removed.

The second type, as shown in Figure 1.9b, is the most widely used USM. The tool (1) vibrates ultrasonically in the direction perpendicular to the machined surface and is fed into the workpiece (3) by a constant force. The abrasive slurry is pumped into the gap between the tool and workpiece, so that abrasive grains (2) act as indenters to generate the brittle fracture in the workpiece material. Circulation of the slurry removes the debris and supplies new abrasive grains.

The input variables of USM are related to machining, slurry, tool, and workpiece material, as follows (Cong and Pei, 2013):

- Static load: about 0.1–30 N
- Ultrasonic vibration frequency: 20–40 kHz
- Ultrasonic vibration amplitude at the tool end: 5–50 mm
- Abrasive type (hardness): BC > SiC > SiO_2 > Al_2O_3
- Abrasive grain size: 15–150 mm
- Abrasive weight concentration: 30–60%
- Tool materials (hardness): nimonic alloy > tungsten carbide > stainless steel > titanium > copper

The main motion is the tool vibration. Its velocity v_t (m/s) can be calculated from the equation (Koryagin et al., 2000):

$$v_t = (4 \cdot f \cdot A)/10^3 \tag{1.7.1}$$

where f is the frequency of vibrations (Hz), A is the amplitude (mm).

The motion of the tool can be longitudinal, perpendicular, or circular, with either the tool or the workpiece rotating. Dependent on feeding motion and tool shape, various kinematic schemes may be performed, similar to EDM.

It can be stated that, in general, MRR increases with rising ultrasonic amplitude and frequency, abrasive concentration, and hardness of the abrasive material. It is also enhanced with reducing viscosity of carrier fluid and tool cross-sectional area. To obtain a lower surface roughness, a decreased abrasive grain size or increased static load, ultrasonic amplitude, and abrasive hardness should be applied (Cong and Pei, 2013). In the case of engineering alumina ceramics, given an amplitude of 0.02 mm, 27 kHz frequency, SiC slurry ratio of 1:1, and static pressure of 2.5 kg/cm^2, the maximum MRR of 18.97 mm^3/min was achieved and surface roughness of 0.76 μm *Ra* was possible (Kang et al., 2006).

Efficiency of the process is proportional to the square fragility criterion, t_x^2, which is the relation of shear strength of a material σ to normal stress applied τ. Materials with $t_x > 2$, such as glass, ceramics, silicon, or germanium, are the most effective when subject to USM (Koryagin et al., 2000).

Devices for USM can be categorized as portable, of smaller power, and stationary, either universal or specialized. Typically, such a device consists of power supply, high-frequency generator, transducer, ultrasonic amplitude transformer (horn), abrasive slurry supply system, tool, tool holder, fixture, and measurement devices for machining depth assessment. Jain et al. (2011) distinguish microultrasonic machining and propose to categorize micro-USM units as stationary, microrotary USM, and hybrid USM. The latter category includes USM + EDM, USM + AFM, and ultrasonic-assisted turning and drilling in microscale.

Rotary ultrasonic machining (RUM) should be mentioned as an important hybrid machining process that utilizes ultrasonic vibrations. This is a mechanical type of nontraditional machining where the basic material removal phenomenon of ultrasonic machining and conventional diamond grinding amalgamates are combined. As a result, a higher material removal rate and an improved hole accuracy with superior surface finish are achieved (Singh and Singhal, 2016). Moreover, the RUM technology has improved the surface quality of the workpiece, increased the processing accuracy, prolonged the cutting tool life, and improved the processing efficiency by up to ten times compared to traditional ultrasonic machining under similar conditions (Zhou et al. 2019).

In the ultrasonic vibration-assisted (UVA) machining processes, as shown in Figure 1.9d, vibration is applied to either the cutting tool or the workpiece in order to reduce the cutting force, cutting temperature, tool wear, cutting chip, and resulting surface roughness of the workpiece, as well as to increase cutting speed. UVA machining modifies both traditional processes, such as UVA turning, UVA drilling, UVA grinding, and nontraditional ones, like UVA electrical discharge machining (EDM) and UVA laser beam machining (LBM) (Cong and Pei, 2013). The UVA machining is most effective for cutting with small MRR, e.g., in threading. When a thread cutting tool is excited with ultrasonic vibrations, momentum can be reduced by 25 to 50% and the resultant thread surface is improved. In effect, numbers of thread cutting tools may be reduced, improving efficiency of the entire process 1.5 and even 3 times while excluding workpiece damage caused by broken taps (Koryagin et al., 2000). Ultrasonic-assisted machining is commonly used for machining of difficult-to-cut materials, especially for ultra-precision machining of titanium alloys. To avoid surface damage and side burrs caused by vibration, a magnetic field can be introduced to ultrasonic-assisted diamond cutting, thus creating a hybrid ultrasonic vibration and magnetic field-assisted diamond cutting process (Yip et al., 2021).

Ultrasonic is also found useful in the ultrasonic surface treatment (UST) process. The UST operation improves tribological properties of a material and fatigue life of engineering systems by enhancing surface mechanical properties, such as roughness and residual stress, by changing the nanostructure induced by severe plastic deformations on the surface (Qian et al., 2016). Ultrasonic shot peening (USSP) treatment is an effective method of fabricating nanocrystalline on the surface of metals. The USSP process involves a generator of ultrasonic signals, a transducer which translates the generated signals into mechanical motion, and a simple metal rod which propels the shots. As a result of USSP, a surface nanocrystalline is formed and hardness is obviously promoted (Zhang et al., 2021). Ultrasonic impact treatment (UIT) can be applied to bulk metallic glasses to obtain tensile yield platforms regardless of zero tensile plasticity of as-cast ones. Surface gradient heterogeneity caused by UIT was found responsible for a favorable shearing path and apparent tensile plasticity (Tu et al., 2021). UST may be combined with other surface treatment methods, e.g., electric spark treatment (EST). It was demonstrated by Lei et al. (2010) that surface roughness and residual stress after the ultrasonic treatment were substantially reduced, and the fatigue life of specimens strengthened by ultrasonic treatments was about twice that of specimens treated by EST only.

It should be emphasized that UST methods can be widely applied to *remanufacturing processes*, especially in combination with other methods. A recent study (Ye et al., 2020) investigated the microstructure and mechanical properties of a Cr–Ni alloy layer deposited by laser cladding (LC) on the surface of a worn shaft made of 1045 steel and then processed by hard turning combined with ultrasonic surface rolling (USR). It was found that USR greatly improved the surface of cladding layer, so that its roughness decreased by 88.5%. Moreover, the residual tensile stress was transformed into residual compressive stress. Microhardness, elastic modulus, and fracture toughness of the cladding layer surface increased significantly after USR.

In order to improve the material removal rate and integrity of a machined surface, hybrid chemical-assisted ultrasonic machining (CUSM) methods are used. CUSM is suitable even for highly demanding materials like bulletproof glass (Singh et al., 2017). Various experiments and comparisons with conventional results of glass machining proved the superiority of the hybrid method, increasing the MRR up to 200%, improving the surface roughness, and decreasing dramatically the machining load (Choi et al., 2007). Application of CUSM can provide process efficiency several dozen times better than EDM and five to six times better than USM, with five times better tool wear resistance and energy consumption reduced by 60–80% (Koryagin et al., 2000). Owing to a higher surface quality and compressive residual stresses in the surface layer, wear resistance and fatigue strength of hard alloy molds, dies, and other components of engineering systems can be improved with application of CUSM.

1.8 ELECTRON BEAM MACHINING

The electron beam machining (EBM) process uses electron beams as energy sources introducing thermal energy in order to remove material from a workpiece by melting and/or vaporization (Liang and Shih, 2016). The kinetic energy of electrons striking the workpiece surface converts into thermal energy of the material, increasing its temperature. EBM has been used in aerospace and nuclear industry since 1960, primarily for welding applications, but at present it is also utilized for microdrilling, cutting, and engraving applications (Bhattacharyya, 2015). Electron beam cutting can also serve fabrication of multiwalled boron nitride nanotubes when a focused electron beam with a diameter much smaller than the tube diameter is used. By controlling the electron beam size, it is possible to cut boron nitride nanotubes and to form sharp, conical crystalline tips (Celik-Aktas et al., 2007).

In EBM processes, a relatively high power density causes heating of the workpiece material only within the spot where the material is removed from by melting and evaporation (Okada, 2019). The ratio between these two phenomena depends on the power density, so that evaporation intensifies as the power density increases. In the EBM process of a low power density, the temperature on the workpiece surface reaches the melting point of a respective material, enlarging the melt pool due to heat conduction, as shown in Figure 1.10a. When the power density is increased, vaporization of the material is intensified and causes voids and keyholes seen in Figure 1.10b. Due to the high pressure of evaporation, the melted material is blown away. Further increases to high power densities of 10^6–10^7 W/cm^2 lead to heating above the boiling point at the spot. The pressure in the keyhole becomes higher than the surface tension of the melting pool so that the material removal effectively progresses along the depth direction, drilling small and deep holes, as illustrated in Figure 1.10c (Okada, 2019).

Thus, *electron beam drilling* involves a deep penetration with high-speed controllability that enables drilling of a large number of micro holes at an extremely high speed. The drillable workpiece thickness is less than several millimeters at its maximum even with repeated pulse irradiations. Therefore, electron beam drilling is applied to drillings of large numbers of holes in difficult-to-drill materials, such as cooling holes on inlet ducts of gas turbine engines, many types of filters, and spinner heads for fiber production. It is possible to drill deep holes into ceramic materials, because the heat conductivity of ceramics is low, so that the high temperature needed for material removal is easy to obtain. It has also been applied to drilling holes for drawing dies made of alumina and diamond (Okada, 2019).

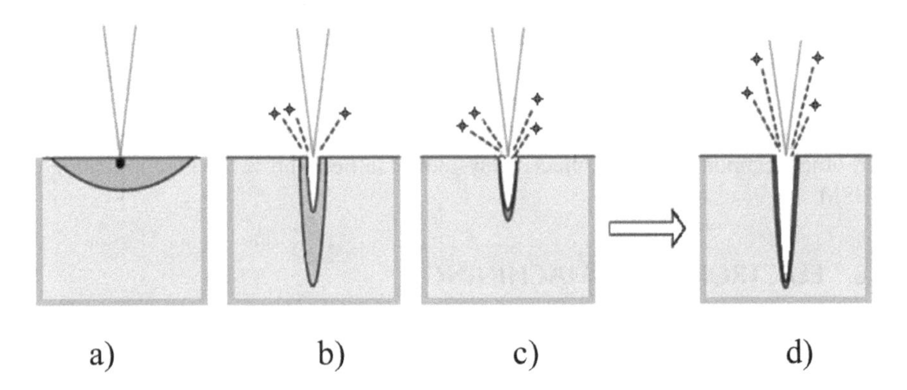

FIGURE 1.10 Material removal mechanisms during electron beam machining at different power densities: *a* – low, *b* – medium, *c* – high power density.

Electron beam welding (EBW) is a fusion welding process based on heat generated by a beam of high-energy electrons, so that edges of a workpiece are fused and joined forming a weld after solidification. EBW is often performed in vacuum conditions to prevent dissipation of the electron beam (Kalaiselvan et al., 2021). EBW is successful in welding of composites, especially SiCp/Al composites. In this material, the huge difference between the properties of SiC particles and matrix alloys causes great difficulties to joining processes. Conventional fusion welding methods are prone to generating defects such as voids and harmful Al_4C_3 phases formed in the reaction between the SiC particles and the matrix in the molten pool, which prevents from meeting the performance requirements of the composite joint (Zuo et al., 2020). Another example of EBW application to composites is TiB_{2p} reinforced aluminum. No obvious pores or cracks were reported in the weld seam with homogenous distribution of TiB_2 particles. Moreover, hardness of the fusion zone and heat-affected zone (HAZ) both increased in comparison with that of the base metal, with no interface reactions between TiB_2 particle and Al matrix (Cui et al., 2010). EBW proved effective at joining SiCp/Al composite with Ti-6Al-4V, welding of Al-Al_2O_3 composites, etc. (Kalaiselvan et al., 2021).

An interesting application of electron beam machining was described by Drobny (2013). EBM can be used to perform the *cross-linking process* improving bonds between individual layers to hold a laminate together. Cross-linking increases cohesive strength of an adhesive and consequently the bond strength in laminates. Laminated materials, opaque to ultraviolet and visible light, may be cross-linked by EB since high-energy electrons penetrate paper, foils, and fabrics. Laminates of thin films or thin-film overlays can be processed by low-energy EBs. For setting adhesive bonds between thicker substrates, higher energy or even X-ray radiation may be used. Materials with very different coefficients of expansion can be bonded by EB-curable adhesives without interfacial stresses that are created when using thermal curing. EB curing is used for curing laminating adhesives in flexible packaging and this application has been growing rapidly since the introduction of low-voltage compact EB processors. Below are listed some other advantages of EB curing of laminating adhesives:

- Adhesives do not require solvents
- An adhesive is one part chemistry (no mixing needed)
- Long shelf-life (more than six months)
- An adhesive remains unchanged until it is cured
- No multiroll coating is needed
- No complex tension controls are needed
- The adhesive bond is established almost instantly
- Real-time quality control
- In-line processing; immediate shipment is possible
- Easy cleanup (Drobny, 2013)

Figure 1.11 shows basic features of an EBM unit. It consists of a pulse energy source (1), electron gun (4) where a high-power electron beam is formed, a vacuum chamber

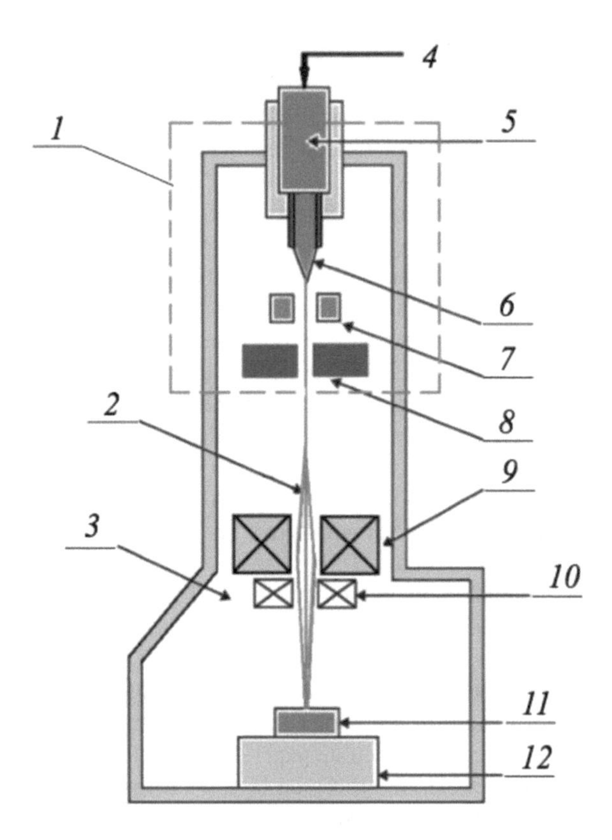

FIGURE 1.11 Electron beam machining principle: 1 – Electron gun, 2 – Electron beam, 3 – Vacuum chamber, 4 – High voltage supply to cathode, 5 – Cathode cartridge, 6 – Cathode filament, 7 – Control electrode, 8 – Anode, 9 – Magnetic lens, 10 – Deflection coils, 11 – Workpiece, 12 – Work table.

where the workpiece (13) is machined, a system of pumps producing vacuum of ca. $1.33 \cdot 10^{-2}$ Pa, a control system responsible for the trajectory of an electron beam, and devices supervising the entire process (Koryagin et al., 2000).

In a vacuum chamber, a tungsten cathode filament is heated to between 2,500 and 3,000°C and is then able to emit electrons under 150 kV. An electron beam is focused with a magnetic or static lens system on the workpiece surface over a well-defined area with a diameter of ca. 0.025 mm. Kinetic energy is converted into heat, causing evaporation of the workpiece material from a specified area due to a high power density to the order of 1.55 MW/mm^2 (Bhattacharyya, 2015). The machining modes are defined by the current I in the beam, accelerating voltage U, energy density q in the focal spot, duration t_p, and frequency f_p of the pulses, as well as by the speed of the spot movement on the workpiece surface. Power of the beam in the pulse can be calculated as follows (Koryagin et al., 2000):

$$P_b = I \cdot U \cdot f \cdot t_p \quad (1.8.1)$$

Thus, both efficiency of the process and the machining depth depend on the power P_b, cross-section area of the beam, and thermal characteristics of a machined material. The main advantages of this process are a high degree of automation, high productivity, high precision, and the ability to machine all types of materials (Bhattacharyya, 2015). However, since static electrical load defocuses the electron beam, conductive materials are preferable.

1.9 ION BEAM MACHINING

Ion beams are useful in such applications as doping of semiconductors or depth profiling in surface spectroscopies. Ion beam techniques are used for surface treatment or depositing coatings on engineering materials to enhance resistance to corrosion and wear of metallic and ceramic materials (Halada and Clayton, 2012). Ion beam machining (IBM) is an important nonconventional manufacturing technology for micro- and nanofabrication. A stream of accelerated ions in a vacuum chamber is able to remove, add, or modify atoms on the surface of an object, mainly as a result of an energetic collision cascade. Transfer of sufficient ion energy and momentum to target atoms forces parts of the ions to be finally implanted into a substrate after losing their energy (Fang and Xu, 2018). IBM does not produce any HAZ as melting and evaporation of workpiece materials does not take place. In effect, the process does not introduce any mechanical strain in the workpiece. Unlike traditional machine tool technologies of cutting, grinding, and lapping, IBM has no inherent reference surface. The process can be used for unusual and difficult tasks like aspherizing of lenses, sharpening of diamond microtone knives and cutting tools, IC pattern etching, etc. Very high cost of an IBM machine makes the process uneconomical (Bhattacharyya, 2015).

IBM can be classified as ion sputtering/etching (remove material), ion sputter coating/ion-induced deposition (add material), and ion implantation (implant modification), according to the function (Fang and Xu, 2018).

Sputtering is a very important process in ion beam machining of micro- and nanostructured devices. Ion milling can be performed using a dedicated system employing a broad-beam ion source to generate a beam of usually Ar ions with beam energy typically in the range 600–1,000 eV (Hindmarch et al., 2012). The ion beam is able to etch away any regions of a thin-film or device heterostructure left unprotected by an overlying resist pattern. Usually, ions sputter and remove atoms near the surface of a target, but a small part of recoil cascade atoms exiting the target might originate from deep inside it (Fang and Xu, 2018).

Ion sputtering efficiency is measured with the number of atoms ejected from the target surface by every incident ion. Generally, higher ion sputter yields can be achieved by choosing heavier ion sources or lower binding-energy target materials, since atom's energy received by collision must be larger than its surface binding energy E_{surf}. High sputtering efficiency can be obtained by using ions with energies ranging from 10 to 100 keV, and the ion energy of up to 30 keV is usually selected in commercial focused ion beams (FIB) (Allen et al. 2009; Xu et al. 2015). Higher energy than the order of 100 keV allows ions to penetrate into the substrate, decreasing the ion sputtering yield or causing ion implantation (Fang and Xu, 2018).

Removal of material on the atomic scale by sputtering is used in *ion beam figuring* (IBF) for production of ultra-precise optical surfaces, such as spheres, aspheres, and free forms on lenses and mirrors. Aside from finishing of optical components, an IBF process can also be employed on molds used to produce optical lenses (Zeuner and Kontke, 2012). Usually, diameters of machined optics can be from 5 up to 2,000 mm, though recently even 1 mm diameters became possible to process. By means of the ion beam etching technology (IBE), feature sizes from < 100 nm up to > 10 μm can be smoothed (Schaefer, 2018). Arnold and Pietag (2015) reported application of ion beam figuring to further reduction of peak-to-valley (PV) sphericity error from the current 50 nm to values less than 10 nm PV with a multi-axis ion beam figuring machine dedicated to deterministic correction of silicon spheres.

Ion beam sputtering, also called *ion beam deposition* (IBD), is a thin-film deposition process that uses an ion source to sputter a target material (metal or dielectric). A typical configuration of an IBD system consists of an ion source, a target, and a substrate. An advanced ion beam source utilizes an electron gun, so the electron beam strikes the surface of a target and produces free ions or clusters. Sputtered species (cations, anions, or neutral particles) cool down when flying around the vacuum chamber and deposit onto a substrate to create either a metallic or dielectric film. Preheating substrates may improve overall performance (Kafle, 2020). When a second ion gun is used to assist the deposition by bombarding the growing film, the method is called *ion beam-assisted deposition* (IBAD). Dual ion beam sputtering can improve adhesion, density, control of stoichiometry, and low optical absorption in thin films, while low-energy reactive ion beam bombardment can increase the rate of compound formation, control the stoichiometry, and improve adhesion of a deposited thin film (Teixeira et al., 2011).

The main advantages of IBD methods are attributed to the fact that the ion beam is monoenergetic, i.e., ions possess equal energy, and highly collimated. As such, it enables extremely precise thickness control and deposition of very dense, high-quality films (Kafle, 2020).

Among novel methods based on the local sputtering of photoresist sidewalls during ion beam etching, ion beam etching redeposition is worth noting (Desbiolles et al., 2019). This method using local *sputter-redeposition* is able to manufacture multi-material 3D nanostructures of various shapes, profiles, heights, thicknesses, and complexity. The process is simple and uses standard microprocessing tools only, but complex nanostructures such as nanochannels, multi-material nanowalls, and suspended networks can be successfully fabricated with features of sub-100-nm dimensions and an unprecedented freedom in material choice. This provides an alternative to traditional nanofabrication techniques, as well as new opportunities for biosensing, nanofluidics, nanophotonics, and nanoelectronics.

Ion implantation is an effective technological tool for introducing single impurities into the surface layer of a substrate in order to modify or change the near-surface chemical composition, affecting surface physical and chemical properties of materials. The degree of material surface modification depends on individual chemical and structural properties and on implantation parameters, such as the type and energy of an implant, current density in the ion beam, and substrate temperature. The most critical parameter is ion dose F_0, measured in ion/cm^2, which determines quantity of an implant, so that ion implantation can be divided into low-dose and high-dose processes (Stepanov, 2019).

Ion implantation techniques introduce elements into target surfaces through implantation of ions that are typically accelerated to energies between 20 and 200 keV. This process operates in a high-vacuum environment and the ions are able to penetrate solid materials up to the depth of several nanometers, which increases as bias voltage rises (Gan and Berndt, 2015). One of the most valuable aspects of ion implantation consists in creating surface layers with enhanced wear or corrosion resistance without significant dimensional changes. For instance, creation of nitrides through nitrogen implantation, near-surface TiC through implantation of Ti followed by C, or ion beam mixing of thin RF-deposited surface layers of Al, Si, Mo, and W into steel, all result in geometric changes of less than 100 nm, yet combined with significant improvements in wear and friction properties (Halada and Clayton, 2012).

Particularly interesting results were obtained on enhancement of commercially available ceramic and cermet cutting tools. The cermet cutting inserts underwent implantation with ions of nitrogen and combination of N^+ with Al^+, which reduced the cutting force tangential component F_c by 15 and 20% and the feeding component F_f by 35 and 40%, while the wear parameter VB_C diminished by 75 and 65%, respectively (Morozow et al., 2018). Implanted with Si^+ and Si^+/N^+, carbide inserts hardness grew by 10%, friction coefficient reduced by ca. 30%, and, during machining tests, exhibited smaller cutting forces and higher wear resistance than inserts without an additional ion implanted layer (Narojczyk et al., 2018). Yttrium and rhenium ion implantation also had impact on tribological properties (Morozow et al., 2019) and cutting performance of nitride ceramic cutting tools (Morozow et al., 2020). Namely, the lowest and the most stable friction force of between 12 and 40 N occurred in IS9 ceramics with yttrium coating ($F_0 = 2 \times 10^{17}$ ion/cm^2) during 800 s of testing, while unimplanted IS9 reached 60 N. Both rhenium and yttrium ions improved wear resistance of IN22 ceramics to a similar degree. VB_N values of wear for implanted tools were at least 20% and at best 75% lower than in the case of non-implanted ones. Cemented tungsten carbide guide pads implanted with 1:1 N_2^+ and N^+ ions at 60 kV acceleration voltage were able to manufacture two times more holes than non-implanted ones (Morozow et al., 2021). At least two mechanisms can be attributed to the increased durability of the nitrogen implanted (WC)-Co guide pads, namely carbon lubrication and nitride strengthening.

1.10 LASER MATERIAL PROCESSING

As emphasized by Shoemaker (2003),

> since the operation of the first laser in 1960, literally hundreds of different laser varieties have been developed and the light that they produce is being used in thousands of applications ranging from precision measurement to materials processing to medicine.

According to Gefvert et al. (2019), almost 60% of laser applications can be categorized as material processing and communications, ca. 30% belong to research, development and military sectors, instrumentation and sensors, medical and aesthetic applications, and lithography, and the rest is divided between displays, optical storage, and printing devices. Dimensional span of laser devices is impressive, too. The smallest laser consists of 44-nm-diameter nanoparticles with a gold core and dye-doped silica shell, with outcoupling of surface plasmon oscillations to photonic modes at a wavelength of 531 nm (Noginov et al., 2009), while the largest occupies a building with a footprint covering an area equivalent to just over three football fields (Spaeth et al., 2016). The most powerful laser beam has been recently launched at Osaka University in Japan, able to produce a beam with a peak power of 2,000 trillion watts (2 petawatts) for an incredibly short duration, approximately a trillionth of a second or one picosecond (Sarri, 2015).

The unique characteristics of laser light make lasers very special devices. These are (Hitz et al., 2001):

1. The capability to produce a narrow beam that can increase its energy intensity so that it is able to cut or weld metal or create tiny and wonderfully precise patterns in a laser printer
2. High spectral purity
3. The way its waves are aligned

Laser processing of materials, both metals and nonmetals, such as cutting, drilling, welding, heat treatment, etc., has several advantages over conventional techniques. Laser drilling can be more quick and less expensive than the mechanical process. There is no tool wear and lasers make cuts with a better edge quality than most mechanical cutters, so that edges of cut metal parts rarely need to be filed or polished. Laser welding can often be more precise and less expensive than conventional methods and is more compatible with robotics. Laser heat treatment involves heating a metal part with laser light, increasing its temperature to a point where its crystal structure changes, causing surface hardening or enhanced wear resistance (Hitz et al., 2001). In the case of new materials like metal foams, laser processing such as laser forming, laser welding, laser cutting, and laser additive manufacturing can be applied successfully (Changdar and Chakraborty, 2021).

1.10.1 Classification of Lasers

According to their active medium, lasers can be categorized as *solid-state*, *liquid*, or *gas* lasers (Powell, 2003). The solid-state laser principle is shown in Figure 1.12.

It is also possible to distinguish five groups: solid-state, semiconductor, liquid, gas, and plasma lasers, as proposed by Šulc and Jelínková (2013). The authors admit, however, that the solid-state and semiconductor lasers can be integrated into one group, because both these active media are solids. In a narrower sense of the term, solid-state lasers are systems whose active medium consists of a transparent solid matrix, e.g., crystal, glass, or ceramics, doped by an optically active ion and using optical pumping for excitation.

Yan and Takayama (2020) emphasize that *solid-state lasers* often use an active species held in an insulating dielectric crystal, glass, or semiconductor material,

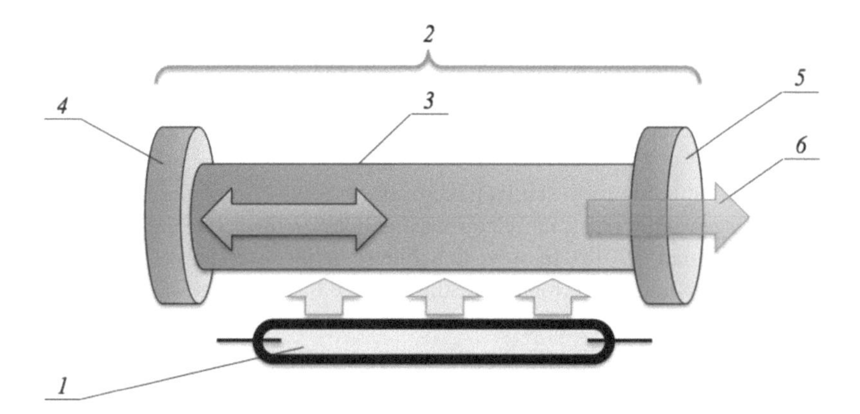

FIGURE 1.12 Operating principle of a solid-state laser: 1 – Flash lamp (pump source), 2 – Optical resonator, 3 – Lasing medium, 4 – Mirror, 5 – Semi-transparent mirror, 6 – Laser output.

where the active species itself is a dopant ion with discrete energy levels. They add that diode pumping has become more common and is expected to completely replace lamp pumping (Yan and Takayama, 2020).

There are many types of optically active dopants, including transition metal ions, rare-earth ions, and color centers. Fundamental properties of a particular laser are determined by a combination of specific optically active point defects and specific host materials. As a result, the operation can be performed by three-level, four-level, and vibronic lasers spanning the near-ultraviolet, visible, and near-infrared spectral regions. The solid-state lasers can be pulsed or continuous wave (CW), discrete line or tunable, mode-locked or operating in Q-switched modes (Powell, 2003).

The best-known solid-state lasers are listed as follows (Abramczyk, 2005):

1. Ruby laser of 694.3 nm, visible region
2. Nd:YAG laser with neodymium-doped yttrium–aluminum garnet matrix (1064 nm)
3. Titanium–sapphire laser (670–1,070 nm) and other solid-state tunable vibronic lasers
4. Rare-earth elements other than neodymium, i.e., holmium, erbium, thulium lasers, such as Ho:YAG, Er:glass, Er:YAG, Tm:YAG, emitting at about 2 μm depending on the matrix and doped material used

Semiconductor laser diodes provide coherent sources of light to an ever-increasing market with applications ranging from the defense industry and medical field to entertainment and fiber-optic communications (Sobiesierski and Smowton, 2016). In 2015–2018, diode lasers accounted for approximately 40% of the world laser market by value (Gefvert et al., 2019).

The fundamental advantage of semiconductor lasers, as compared to most other laser types, consists in operation using direct current injection to provide population inversion. In effect, a high electrical to the optical power conversion efficiency of up to 75% is reached, as well as high reliability, compactness, and low manufacturing costs (Sobiesierski and Smowton, 2016). A semiconductor laser is very compact, typically, the size of an active laser part is only 100 μm × 1,000 μm × 100 μm (Razeghi, 2010).

The optical emission from semiconductor lasers arises from a radiative recombination of charge carrier pairs, i.e., electrons and holes in the active area of the device (Koch and Hoffmann, 2018). In order to achieve carrier inversion, it is necessary to excite sufficiently many electrons from the valence band into the conduction band, which is referred to as pumping the semiconductor laser. This procedure can be executed optically, but the possibility of electrical pumping with a few tens of milliamperes at voltages of a few volts can be considered another advantage of semiconductor lasers (Koch and Hoffmann, 2018). Moreover, output of a semiconductor laser can be easily modulated by modulating its injection current, which further expands the range of its applications (Dutta, 2003).

In the semiconductor structure, there are two types of bands: conduction bands, which consist of unoccupied states; and valence bands, which consist of occupied states. Most III–V and II–VI compounds (the numerals refer to columns in the Periodic Table) are direct bandgap materials, where the conduction-band energy minimum and the valence-band energy maximum have the same momentum. GaAs is an example of a direct bandgap semiconductor. On the other hand, if the band extrema occur at different momentum values, the semiconductor has an indirect bandgap, e.g., SiGe (both column IV) and AlAs (III–V). In general, a semiconductor electronic band structure has numerous bands with asymmetric shapes and sometimes several energy maxima and minima (Chow and Koch, 1999). A large variety of semiconductor materials are suitable for semiconductor lasers. For example, materials covering the red are (AlGaIn)P, the near-infrared (AlGa)As or (GaIn)As, the telecom range (GaIn)(AsP), infrared (GaIn)(AsSb), and even the blue-ultraviolet range (GaIn)N. The most stringent requirement for a material is that it has a direct bandgap, which tends to have a high radiation transition rate. This condition excludes the elemental semiconductors, silicon and germanium, from use in semiconductor lasers (Koch and Hoffmann, 2018).

Initially, the semiconductor laser was built with a simple *p-n* junction using the same material for the active and surrounding layers, which is referred to as a homojunction. To increase laser efficiency, multiple layers with different optical properties were used in the laser structure, forming heterojunction lasers. A further improvement can be obtained by sandwiching an active GaAs layer between two AlGaAs layers. This structure is called a double-heterojunction (DH) or double-heterostructure To date, the most extensively used heterostructure semiconductor lasers are in the GaAs-AlGaAs and GaAs-InGaAsP systems (Razeghi, 2010).

Color center lasers (CCL). Color centers are lattice vacancy defects trapping electrons or holes, usually created in single crystals at room temperature under ionizing radiation (Courrol, 2006). Color centers in alkali halide crystals can be applied as high-gain active materials in tunable solid-state lasers, cryogenically cooled and optically pumped. CCLs have low threshold pump powers, relatively high output powers, and broad emission bands. With a large variety of color center types and host lattices, the combined tuning range covers the near-infrared region from about 0.8 to 4 μm (Gellermann, 1991). Color centers exist in many types of crystalline solids, but most promising materials, apart from alkali halides, are alkali-earth fluorides, oxides (e.g., Al_2O_3), and covalent crystals, with diamond as the only available representative. Among the alkali halides, LiF and NaF are the most outstanding because of their low hygroscopicity, while LiF is the most durable and suitable for high average-power laser operations at room temperature. Sapphire and diamond feature a unique combination of thermal and mechanical properties, but they are not as easily accessible for technology as alkali halides (Basilev et al., 2003).

Another important group of solid-state lasers, worthy of distinction as a self-standing class, consists of *high-power fiber lasers*. In fiber lasers, active (laser-gain) optical fiber is combined with one or more pump lasers, usually laser diodes. There are various types of fiber lasers, including low-power continuous wave, low- and high-energy pulsed, etc. Fiber lasers are relatively simple in design and easy to maintain. Since they are pumped with laser diodes, these lasers are also rugged and long-lived. Kilowatt-class CW fiber lasers find common application in materials processing, such as cutting, welding, brazing, and surface treatment (Wallace, 2016). Among benefits important for industrial purposes, the following can be listed: high power with low beam divergence, flexible beam delivery, low maintenance costs, high efficiency, and compact size (Quintino et al., 2007). Sadikov and Mogilevets (2016) point out the following advantages of high-power fiber lasers used in industrial materials processing:

1. Long life of above 100,000 working hours can be further expanded at reasonable expenses, maintenance costs close to zero.
2. Very short time and low expenditure required to prepare the location and to launch these devices.
3. Universal laser source without specific technical features, which can be reoriented from one technological process to another.
4. It is possible to increase the power of a laser. For instance, having a 700 W device, one can buy pumping blocks and obtain up to 2,400 W without any additional rebuilding. These additional blocks can be installed in two or three hours.
5. Transmission of irradiation with a 10- to 100-m long fiber simplifies the design and composition of manufacturing and materials processing systems. Application of a wide range of industrial robots is possible.
6. Fiber lasers offer prospects of organizing multipurpose and multifunctional technological areas to increase the productivity of the laser source.
7. In terms of staff, there is no need to employ additional specialists to maintain the laser system. An operator can be trained in a week's time, so that existing staff may enter a new, higher productivity level.

Another large group of lasers is based on a gaseous active medium. The working principle of the *gas lasers* is very similar to that of solid lasers, but the energy of emitted light depends on a gas mixture excited by a high voltage electrical discharge, typically 2–4 kV. The mixture normally consists of an inert gas that absorbs the discharge energy and transfers it to an active gas atom, whose acceptable energy levels determine the emission energy. A large variety of gas mixtures are used in lasers and power conversion efficiency can range from 0.01 to 15% (Razeghi, 2010). Lasers with a gaseous active medium are advantageous with their wide tunability, high flexibility, relatively low cost, beam quality, and power scalability. However, the recent popularity of semiconductor lasers seems to have overshadowed gas lasers (Endo and Walter, 2006).

Excimer lasers are pulsed gas lasers with an efficient and powerful broadband emission in several spectral ultraviolet regions with typical spectral widths of ca. 2 nm (Sze and Harris, 1995). An exception to this categorization is the XeF laser with its broadly tunable C→A transition (approximately 50 nm) in the visible spectrum. Two basic formation channels for the excited state can be named: (1) recombinations of positive rare gas ions with halide ions, and (2) reactions of excited rare gas atoms with halogen compounds. The laser excitation techniques are primarily as follows: high-energy electron beams, electron beam sustained discharge, neutron pumping from reactors, preionized avalanche discharges, and microwave excitation. The pulsed electron beam and preionized avalanche discharge techniques appear the most useful. Among the best-known excimer lasers are ArF, KrF, XeCl, and XeF (Sze and Harris, 1995).

Chemical lasers use a pumping method based on energy released during a chemical reaction. The reactions most commonly used in commercially available chemical lasers consist in the generation of excited HF* and DF* molecules. HF and DF lasers can produce radiation up to several hundreds of watts in the far-infrared and infrared regions (1.3–11 μm) as well as in the visible range (Abramczyk, 2005). The most efficient high-energy chemical laser with the shortest wavelength is the chemical oxygen-iodine laser (COIL). It operates on a near-infrared radiation of atomic iodine, $\lambda = 1.315$ μm. The laser medium, I*, is produced through a pumping reaction between singlet oxygen and ground state iodine (Hu et al., 2007). Comparing the lasing power for transonic and supersonic injection schemes, the output power and chemical efficiency for the latter were about 20% higher than for the transonic mixing scheme (Barmashenko et al., 2003).

The main component of a free-electron laser (FEL) is an undulator magnet that forces a large number of electrons to emit their radiation coherently. Both the role of the active laser medium and the energy pump are taken over by a relativistic electron beam. The FEL radiation is almost monochromatic and well collimated. Among the important advantages of the FEL is the free tunability of the wavelength by simply changing the electron energy (Schmüser, 2003). Other benefits are as follows (National Research Council, 1994):

- High peak power up to gigawatt order is achievable.
- Flexible pulse structure. Not only picosecond pulses with sub-picosecond jitter can be produced, but also the interval between pulses can be varied, offering the option of generating complicated pulse structures.

- FELs easily achieve desirable laser characteristics, such as a single transverse mode, high spatial and temporal coherence, and flexible polarization properties.
- Broad wavelength coverage with the shortest achieved wavelength 240 nm.

However, FEL is usually large and its cost remains high, and relatively little effort has been undertaken to produce smaller and less expensive devices. Moreover, their utilization in scientific research and operation of the facility involve high costs of maintenance (National Research Council, 1994).

Descriptions of other lasers and their working principles can be found in Webb and Jones (2003).

1.10.2 Laser Beam Machining Methods

Laser beam machining (LBM) is recognized as suitable for the processing of difficult-to-machine materials, such as ceramics, titanium alloys, or nickel-based alloys, because of its independence from material hardness, brittleness, and other properties. The material removal mechanism is based on an extremely high power density of the laser beam focused on a very small area. Some of the light is reflected back, but the remaining energy is absorbed by the material surface layer and conducted to the lattice by means of the photoelectric effect. As a result of energy absorption, surface atoms of the workpiece get excited, raising the temperature to material vaporization and melting. Schematically, the mechanism of material removal can be described as energy absorption – a rapid rise in temperature – melting and evaporation – material removal. The latter stage may be also gas assisted (Nagimova and Perveen, 2019). The principle of LBM is shown in Figure 1.13.

Among the several types of LBM, three variations can be classified according to the number of dimensions:

1. 1D – drilling
2. 2D – cutting
3. 3D – milling, turning, grooving, and micromachining (Nagimova and Perveen, 2019)

Laser drilling is a popular nonconventional microdrilling technique. Its main advantages are high processing speed, high efficiency, localized processing, high precision, cost-effectiveness, and no tool loss (Ren et al., 2021). There are four ways of laser drilling distinguished, namely, single pulse drilling, percussion drilling, trepanning, and helical trepanning. In the *single pulse drilling*, a high-energy laser pulse is used to drill either a through hole of diameter below 1 mm in a thin sheet or a shallow blind hole in a thick plate. *Percussion drilling* applies a series of identical laser pulses of short duration, directed on the same spot at a high speed. Each laser pulse removes some amount of material down to a certain depth, obtaining a through hole with less taper in a relatively thick plate. *Laser trepanning* allows for producing large holes by drilling a series of overlapping holes around their perimeter, forming

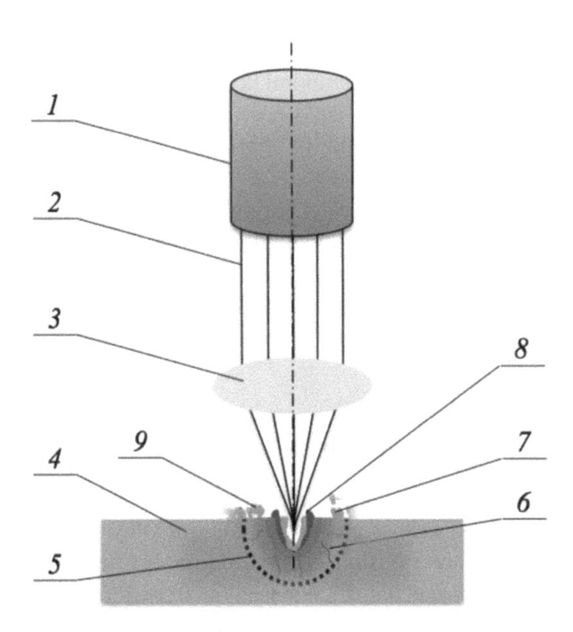

FIGURE 1.13 Laser beam machining principle: 1 – Laser discharge tube, 2 – Laser parallel beam, 3 – Lens, 4 – Workpiece, 5 – Heat-affected zone (HAZ), 6 – Microcracks, 7 – Molten material, 8 – Recast layer, 9 – Surface debris.

a sort of cutting process with laser to cut a contour out of the plate. Thus, holes of any size and shape, both circular and noncircular, can be produced at a constant laser intensity and with an improved quality. In *helical trepanning*, a focused laser beam is rotated around the perimeter lowering its focal position with each rotation and thus gradually deepening the hole (Nath, 2014).

The obtainable peak power density is dependent on the pulse durations. Thus, for millisecond laser drilling, it is reported to range 106–109 W/cm^2, for nanosecond laser drilling in the range of 1–10 GW/cm^2, and for ultrafast picosecond lasers used for drilling holes, in the range of 10–1000 GW/cm^2, decreasing thermal effect on the zone drilled to very small values close to zero (Xia et al. 2021). Accordingly, the quality and accuracy of holes are improved as the duration of the laser pulse becomes shorter (Nath, 2014).

Further improvement of laser drilling efficiency and quality can be obtained by means of magnetic field-assisted technique with or without assist-gas (Wang, Xu et al., 2019), water assistance (Ren et al., 2020), ultrasonic assistance (Xia et al., 2021), electrochemical corrosion assistance (Duan et al., 2020), etc. For example, in the case of femtosecond laser drilling in alumina ceramics, the effect was considerably increased by water assistance, with the taper angle diminishing by 64.19% to 4.20° and the hole cross-section area increasing by 46% (Ren et al., 2020).

Laser cutting is a well-established technology that is nevertheless rapidly developing due to its excellent performance in terms of cutting speed, good yield and width of a cut during machining of difficult-to-cut materials, especially metals and advanced composites in many mechanical industries (Boujelbene et al., 2021). It can be applied where remote control is required for dismantling tools due to dangerous environmental conditions. For instance, underwater laser-cutting techniques with high-power fiber lasers proved to be feasible, producing less secondary waste than other approaches, such as the waterjet and plasma cutting techniques (Oh et al., 2021).

Laser cutting of sheet metal is usually performed from a CW laser source with a coaxial flow of high-pressure assist-gas that removes molten material. Pulsed laser materials processing exploits rapid deposition of laser energy, limiting heat conduction losses compared to CW exposure (Lutey et al., 2018). *Laser cutting of metals* is used for forming and precise processing of shaped parts, but different materials require some modifications of the cutting process. Mild steels can be laser-cut with a superior edge quality compared to conventional technologies both in thin sheets and in thick plates. Stainless or chrome-nickel steel is often laser-cut with nitrogen assist-gas, but the maximum thickness is less than that for mild steels at equal laser power and similar cut-edge quality. Aluminum is second after steel in terms of the volume of industrial laser-cutting applications. Because of its quick oxidation, Al is often laser-cut with nitrogen or air assist-gas. Titanium is usually laser-cut with an inert assist-gas, e.g., argon, which enables finer cutting with microcracks eliminated. Laser can cut both pure copper and copper alloys, such as bronze and brass, but bronze is not often laser-cut because it is usually formed from castings. Oxygen-assisted laser cutting of copper alloys can be performed at higher cutting speeds than the neutral or inert gas-assisted process (Caristan, 2004). For sheet-metal cutting operations, process planning and production planning can nowadays be automated to some extent, so that 2D cutting operations focus on minimization of waste material (Verlinden et al., 2007). Laser-cutting techniques are largely standardized so that various companies throughout the world achieve similar results in metal sheet cutting in terms of efficiency, thickness, or cutting speeds.

The *laser welding* process is not only easily automated, but various real-time monitoring technologies can be applied for improving welding efficiency and guaranteeing the quality of joint products. An extensive review provided by Cai, Wang et al. (2020) distinguishes three different parts of the monitoring: pre-processing scanning, in-process monitoring, and post-process diagnosing. The pre-process scanning mainly focuses on the joint gap between workpieces and seam tracking problems to ensure the central position of the laser beam spot in the gap and thus to obtain reliable joints. The real-time in-process monitoring is concentrated on welding zone characteristics such as keyhole, molten pool, plasma, and spatters. Based on the analysis of dynamic changes of these characteristics, the quality of a weld seam can be adjusted using AI-based methods. The post-process diagnostics is focused on defects like pores, cracks, spatters, surface collapses, underfills, etc., which are critical indicators of the weld seam quality. Among the various sensing techniques, the authors present acoustic emission measurement, optical signal, and thermal signal. Novel monitoring methods, such as X-ray imaging, in-line coherent imaging, and magnetooptical imaging were demonstrated to achieve excellent results. However, a welding situation can be reflected more effectively and comprehensively when multi-sensor fusion technology is applied and full advantage is taken of various signal sensors. Vision sensors are at the core of this technology, so that in an established monitoring system, an optical sensor can be combined with a vision sensor, an X-ray system with a high-speed camera, a sound sensor assembled with a vision sensor, etc. In order to achieve various monitoring objectives, such as parameter optimization, feature prediction, seam tracking, defects classification, simulation validation or adaptive control, a variety of AI techniques can be applied (Cai, 2020). Laser beam welding is a high-quality fusion joining process that enables the welding of dissimilar components, e.g., titanium alloys with various counterparts including steel, aluminum, magnesium, nickel, niobium, copper, etc. (Quazi et al., 2020). It has also been reported that femtosecond laser pulses at high repetition rates can be used to weld glasses of different combinations (Richter, 2016).

Laser marking and engraving machines experience increasing popularity across end-user verticals due to the enhanced performance of laser markers over traditional material marking techniques. As far as end users are concerned, they are applied primarily in machine tools sector, semiconductor and electronics, automotive, medical and healthcare, aerospace and defense, and packaging. The machine tools vertical is estimated to account for the largest share of the laser marking market in 2018, which can be attributed to the ability of laser markers to provide permanent alphanumeric details on objects in terms of product and brand names, batch numbers, manufacturer codes, 1D and 2D barcodes, logos, designs, manufacturer codes, dates of manufacturing, product-related information, and other details (Laser Marking Market, 2021). The most widely used types of lasers in this application are fiber lasers, diode lasers, solid-state lasers, and CO_2 lasers, respectively.

Lasers also demonstrate their feasibility for *micro- and nanomachining*, i.e., fabrication of components or products with at least one feature size in the micrometer or nanometer scale. Laser systems integrated into multi-axis micromachining systems can be used for microscale drilling, cutting, milling, and surface texturing. This way, micro-components made of different kinds of workpiece materials, such as metals, polymers, glasses, and ceramics, may be processed. In particular, the importance of *laser micro-milling* as a micromanufacturing technology is increasing in rapid prototyping, component miniaturization, and batch fabrication methods of serial production of micro-devices (Gao and Huang, 2017).

There are two basic approaches to *laser micromachining*, namely, mask projection and direct-write. Mask projection is normally performed with excimer lasers and with a normal binary mask of 0% or 100% transmission. A material is removed to the same depth in all exposed regions. Stepped multilevel structures can be produced using a sequence of exposures with different static masks and variable height surfaces can be obtained by mask- or workpiece-dragging or by static projection using a half-tone mask. For the direct-write approach, primarily solid-state lasers are more appropriate. In this case, a focused laser spot follows a predefined tool path on the workpiece surface, which is particularly well suited to prototyping because it does not require a mask (Hocheng et al., 2014).

Both the approaches are also applicable to *laser nanomachining*. For instance, a particle mask can be formed by depositing a monolayer of microspheres on the substrate surface and nanoscale pits on a glass surface can be machined with a femtosecond laser (Zhou et al., 2007). Femtosecond laser direct writing has been recently recognized as a promising nanomachining technique able to solve problems of 3D architecting, otherwise impossible. It demonstrates a unique three-dimensional processing capability, arbitrary-shape designability, and high fabricating accuracy up to tens of nanometers, far beyond the optical diffraction, that are desirable in various scientific and industrial fields (Zhang et al., 2010).

Potential of the laser micro- and nanomachining is closely connected with development of microelectronic industry and increasing investment in healthcare technologies.

1.10.3 Laser Surface Treatment

Laser surface treatment of materials has become an important technique with a potential to enhance various properties of different material surface layer, such as surface strength, hardness, roughness, coefficient of friction, and chemical and corrosion resistance (Shukla and Lawrence, 2015). Such improvements are good solutions not only for applications under high wear rates and shear stresses but could also be used for maintaining or elongating a component's functional life by means of covering microcracks in surfaces, e.g., in technical ceramic-based components. In addition, appearance can also be improved, especially in the case of ceramics, by creating a modified surface layer (Shukla and Lawrence, 2015).

Laser surface treatment is a thermal process superior to conventional furnace heat treatment. It is based on heating caused by light adsorption of a surface layer and cooling ensured by high conductivity of a material. The adsorbed laser energy results in a thin surface layer with desired properties while the bulk of the material is unaffected (Muresan, 2015). Commonly recognized advantages of laser surfacing are as follows (Steen, 2003):

- Chemical cleanliness
- Controlled thermal penetration and therefore distortion
- Controlled thermal profile and therefore shape and location of a heat-affected region
- Less after-machining, if any, is required
- Remote non-contact processing is usually possible
- Relatively easy to automate

Application of lasers to surface treatment includes the following (Steen, 2003):

- Surface heating for transformation hardening or annealing
- Surface melting for homogenization, microstructure refinement generation of rapid solidification structures and surface sealing
- Surface alloying for improvement of corrosion, wear, or cosmetic properties
- Surface cladding for similar reasons as well as changing thermal properties such as melting point or thermal conductivity

- Surface texturing for improved paint appearance
- Plating by Laser Chemical Vapor Deposition (LCVD)
- Laser Physical Vapor Deposition (LPVD)

Among the *laser cladding* routes are those which melt a preplaced or blown powder and those which decompose vapor by pyrolysis. In LCVD, the process is based on photolysis, while in LPVD upon local vaporization. Other methods employ sputtering, enhanced electroplating, or cementation. The particle injection process is similar to laser cladding by the blown powder route, but the particles blown into the laser melt pool remain solid. The as-created structure displays improved hardness and wear resistance with reduced friction coefficients. In general, process variations can be introduced to the particle delivery system, pressure delivery, and the gas shrouding systems (Steen, 2003).

Besides the methods that apply laser irradiation to cladding or powder alloying, the laser-induced nitrogen and carbon uptake is an established method to obtain enhanced properties of alloys and their surfaces (Höche et al., 2015). The process of *laser nitriding and carburizing* is a very complex interaction difficult to describe in detail. The entire process is dominated by local heating and the resulting surface temperature, which determines the uptake of nitrogen and carbon. Depending on melting or evaporation effects, different process chains are involved determining coatings and their properties in an interlinking way. Each subprocess causes consequences for the resulting synthesis and has to be examined and weighed in order to understand and to control the process. Just to mention the most important (Höche et al., 2015):

- Laser absorption and local heating
- Melt and evaporation processes
- Plasma expansion into the background gas
- Dissociation and/or ionization
- Gas adsorption and absorption
- Gas atomic transport (diffusion, convection)
- Nucleation and solidification
- Solid-state phase transformation

Attempts at the production of TiN-coatings in reactive atmospheres by means of laser irradiation were undertaken in the 1980s and 1990s, providing an enormous improvement to surface properties (Höche et al., 2015). With a CO_2 laser, it was possible to obtain a hundred micrometers thick coatings. The surface of commercially pure titanium (cp-Ti) after laser nitriding contains a mixture of α'-Ti and δ-TiN. When nitrogen content of the processing gas is increased, the volume fraction of δ-TiN rises at the expense of the α'-Ti which causes surface hardness to grow. Laser nitriding of high-strength (α+β)-Ti alloy Ti-6Al-4V can increase hardness to moderate values in the range of 400–600 HV, improving cavitation and water droplet erosion resistance. Values above 600 HV could significantly improve sliding wear resistance, while abrasive wear resistance of Ti–6Al–4V can only be improved if thick layers exhibiting hardness above 800 HV are generated. Similarly, (α+β)-Ti alloys like Ti-4, 5Al–3V–2Mo–2Fe (SP700), and β-Ti alloys like Ti–10V–2Fe–3Al can be enhanced by laser nitriding (Höche et al., 2015).

Laser carburizing was initially performed using graphite coatings in laser surface hardening of steels. At present, laser carburizing is applied to commercially pure iron, plain carbon, and low-alloy steels. It can be achieved by two different mechanisms: (a) laser surface alloying, which involves melting of a surface layer while carbon enters the liquid phase; and (b) solid-state diffusion of carbon, activated by laser heating (Katsamas and Haidemenopoulos, 2001). Laser carburization also proves successful for austenitic stainless steel and non-ferrous materials such as aluminum and silicon in methane atmospheres (Höche et al., 2015).

Laser surface treatment also has an impact on technical grade ceramics, e.g., Si_3N_4 or ZrO_2, causing changes in their topographical, chemical, microstructural, compositional, mechanical, and thermal properties. Changes in hardness and the resulting crack length from diamond indentation demonstrate modification of the fracture toughness K_{Ic} after fiber laser engineering of surfaces of these two ceramics (Shukla and Lawrence, 2015).

Laser is also used for supporting other surface engineering techniques, for example, enhanced electroplating. In this method, irradiation of a laser beam on a substrate (cathode) during electrolysis promotes drastic modification of the electrodeposition process in the irradiated region. Other applications of lasers to surface engineering are laser cleaning, paint stripping or laser surface roughening. The latter, executed with pulses from an excimer laser, improves adhesion of glue to a surface. Laser shock hardening or "laser shot peening" has emerged as an industrial process able to create a compressive stress in a surface and thus increase fatigue strength of a component's material (Steen, 2003).

1.10.4 Remanufacturing with Laser Cladding Technology

Even though laser cladding remanufacturing (LCR) has been researched actively for several decades and proved to be an effective repair technology, the quality and repeatability of cladding layers are still the key issues in the LCR technology. There are some unquestionable benefits of LCR, but also some important challenges which can be found in the literature (Liu et al., 2017). Among the advantages, the following should be mentioned:

- LCR is applied to both surface engineering and three-dimensional deposition forming.
- Inherent rapid heating and cooling during the LCR process improve the microstructure and property of cladding layers.
- Sound metallurgical bonds between an added metal and a base material are achieved.
- LCR produces very few microcracks and distortions.
- The LCR technique has the ability to manufacture multifunctional homogenous or heterogeneous structures.

The main technical challenges of LCR include:

- Low repeatability and quality of repair layers
- It is difficult to achieve adaptive control
- Difficult and expensive post-process inspection
- Accumulative error

Through optimization of the laser cladding parameters, it is possible to produce smooth, crack-free, and low porosity rate Ni60A clad layers which are metallurgically bonded to a 45 steel substrate (Wu, Li et al., 2019). The authors reported that linear energy density served to identify a threshold for the elimination of large porosity and statistical analysis helped to optimize process parameters. It was demonstrated that powder feed rate in the coating process was the most significant parameter considering the porosity area and cracks were effectively eliminated by a combination of preheating to 300°C and placing an insulated plank under the substrate. Coating of a crack-free large and low porosity of 0.027% were obtained successfully. Its average microhardness was about 2.7 times higher than that of the substrate.

In the case of Inconel 718 and other nickel-based superalloys that are generally used in aerospace, gas turbines, turbochargers, etc., directed energy deposition techniques provide a cost-effective solution for repairing highly valuable components and reduce the turnaround time for fabrication and replacement with a new component (Shrivastava et al., 2021). However, component parts deposited near the edges show the presence of pores which affect the strength and functionality of the component. Moreover, alloy microstructure is sensitive to the process parameters causing considerable variations in the microstructure of deposition layers in the +z direction.

The following can be mentioned among recently reported achievements of LCR: repair of turbine blades from Ni-based superalloy CMSX-4 through the formation of monocrystalline CMSX-4, repair of tools used in soil cultivation to enhance wear resistance due to formation of intermetallic compounds Stellite-6/WC on a B27 boron steel substrate, upgrades to a barrel-screw system Ni40 and Ni60 on C60 steel used in plastic injection molding to improve microhardness, repair of molds and dies employed in hot and cold working through cladding CPM9V steel on an H13 tool steel substrate, repair of railway wheels to increase the hardness of clad materials and reduce wear, or replacement of natural bone, possible through cladding titanium hydroxylapatite on a nitinol substrate, since hydroxyapatite coating reduces nickel release (Siddiqui and Dubey, 2021).

Environmental impacts of remanufacturing can be understood in the example of a cast iron cylinder head block. The analysis performed by Liu et al. (2016) compared remanufacturing through laser cladding with new cylinder head block manufacturing. Considering resource and energy consumptions of these processes and based on six selected environmental impact categories, it was demonstrated that cylinder head remanufacturing by laser cladding would achieve large environmental benefits, cutting environment impact over the entire life cycle by 63.8% on average.

1.10.5 Future Development of Laser Applications

Various laser and laser-assisted techniques prove useful in the restoration of worn machine components, especially in motor industrial and aircraft applications (Tsegelnik, 2015). Laser cleaning is widely applied to the restoration of aircraft turbines, turning surface contaminations to gas or dust fractions. One of the main benefits of this method is the possibility of cleaning without disassembling the aircraft engine, thus decreasing time consumption. Similarly, damaged paints and varnishes can be easily and successfully removed from aircraft surfaces. In the advanced robotic laser coating removal system, a group of mobile walking robots can perform successfully unmanned cleaning of smaller aircraft outer surfaces (Tsegelnik, 2015).

Prospects of further development of laser material processing technologies can be obviously derived from the undisputable advantages of lasers over traditional methods (Haranzhevsky and Krivilev, 2011):

- Extremely high spatial localization of energy
- Exceptionally short time of action, setting important limits to dimensions of a heat-affected zone
- Lack of contact
- Strict dosage of energy

Despite the large variety of existing lasers, radically new laser systems still continue to emerge, such as a quantum cascade laser (Pecharromán-Gallego, 2017). Of course, numbers of newly developed basic lasers have decreased in recent decades, but the broad range of applications is well established and widely applied throughout the world. Moreover, there is still a growing number of possible new applications of the existing laser technologies. Capabilities of lasers are multiplied by an immense number of available wavelength combinations, pulse durations, shapes and power levels, thus expanding areas of potential future applications. This range is even more extensive if possible combinations of lasers with other material processing methods are considered that provide an additional large number of hybrid technologies.

1.11 PROCESSING WITH PLASMA

Plasma is a collection of free atoms or molecules which is partially or fully ionized and which is normally charged neutral, because each particle interacts simultaneously with many others, exhibiting a "collective behavior" (Stevens, 2000). Partially ionized plasmas have electron and ion densities in the range of 10^{15}–10^{19} m^{-3}, neutral species densities in the range of 10^{19}–10^{22} m^{-3}, and pressures ranging from 0.133 to 1330 Pa. The combination of electrical, thermal, and chemical properties of these plasmas makes them uniquely feasible for material processing, for the following reasons (Stevens, 2000):

- Electrons in processing plasmas are not in thermal equilibrium with neutral ions or chamber walls, reaching high temperatures of up to 3000 K. The higher electron temperatures in plasmas produce enhanced chemical

reaction rates and also allow chemical reactions to occur, impossible in other conditions.
- Ions can be drawn out from a plasma surface at energies of tens to hundreds of eV, making possible anisotropic etching and deposition.
- Due to the low density of charged plasma species, plasma particles can interact with a surface without heating it significantly.
- Low-pressure operation allows plasmas to utilize process reactants with high efficiency, which reduces waste and pollution.

Basically, thermal plasma jets can be generated in devices called plasmatrons by direct current (DC), alternating current (AC), radio frequency (RF), or other discharges (Cao et al., 2016). Among DC methods, two main types are distinguished, namely, free-burning arcs (transferred arcs) and plasma torches (non-transferred arcs), which are widely used in materials processing (steel production, metal cutting and coating). The basic design of a plasma torch consists of two electrodes and a cylindrical discharge chamber where a plasma-forming gas is introduced under pressure. In general, there is a button-type cathode, a gas distributor, and a nozzle anode which forms a discharge chamber and usually acts as an arc constrictor in non-transferred DC plasma torches. The arc is initiated between the cathode and the anode by a high voltage discharge, while the plasma-forming gas is supplied through the gas distributor radially or tangentially. Typically, the cathode consists of a water-cooled copper holder with a press-fitted rod of alloyed tungsten, hafnium, or graphite (Mostaghimi et al., 2017). In free-burning arcs (transferred torches), a treated conductive material serves as an electrode, so that the plasma jet is formed in the arc column, as shown in Figure 1.14 (Dragobetsky et al., 2012).

Variations of DC plasmatron designs include axial and coaxial plasma generators, designs with toroidal electrodes, with outer plasma arcs, and with eroding electrodes (Gevorkyan et al., 2016). Among AC plasma generators, high-frequency and ultrashort-wave dischargers find a wide range of applications (Wegman et al., 2010), as shown in Figure 1.15.

High-frequency capacitive (HFC) plasmatrons work in the range of frequencies between 10 and 50 MHz (Toumanov, 2003). The value of their discharge current is limited due to the resistance of the capacitive coupling, so that operational frequency should not be lower than 10 MHz. The specific intensity of the electric constituent of the electromagnetic field is in the range of 100 to 400 V/cm, the voltage on electrodes is within 5–15 kV, and the currents are within 3–15 A. The permissible range of operation frequencies lies between 13.56 and 27.12 MHz. An important feature of HFC discharges is a low value of minimal power necessary for sustaining the discharge, so that the range of practically assimilated powers is 1–100 kW. In a *high-frequency inductive (HFI) plasmatron*, a discharge chamber (5) is a tube made of a dielectric material placed in a coil (4), as shown in Figure 1.15a. The magnetic field is oscillating due to a high-frequency current in the coil and induces a vortex electric field, which in turn ignites and sustains a discharge. HFI plasmatrons can vary in design: some include one grounded end of the coil, some a coil supplied according

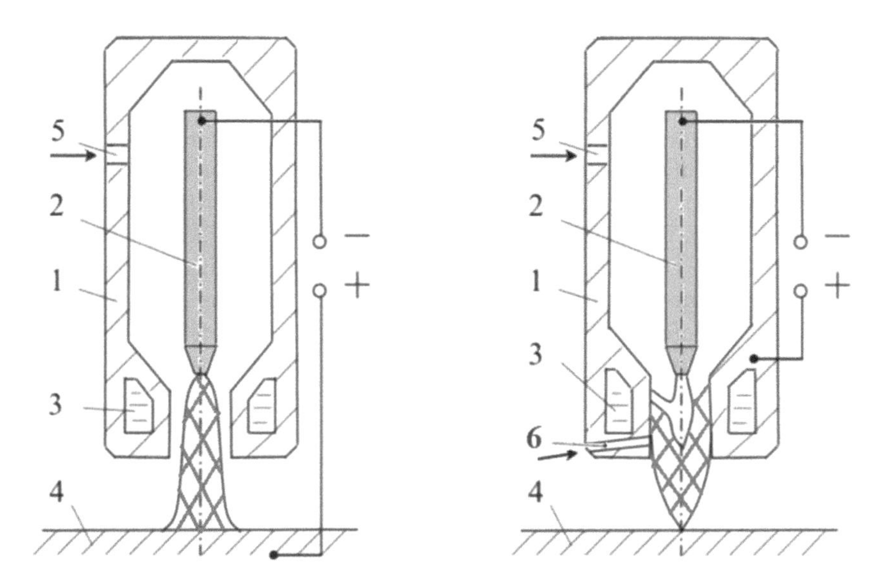

FIGURE 1.14 Direct current plasmatrons: (a) free-burning arc (transferred arc), and (b) plasma torch (non-transferred arc). 1 – Nozzle, 2 – Tungsten electrode, 3 – Coolant, 4 – Workpiece, 5 – Injection of plasma-forming gas, 6 – Channel for the introduction of a processing powder.

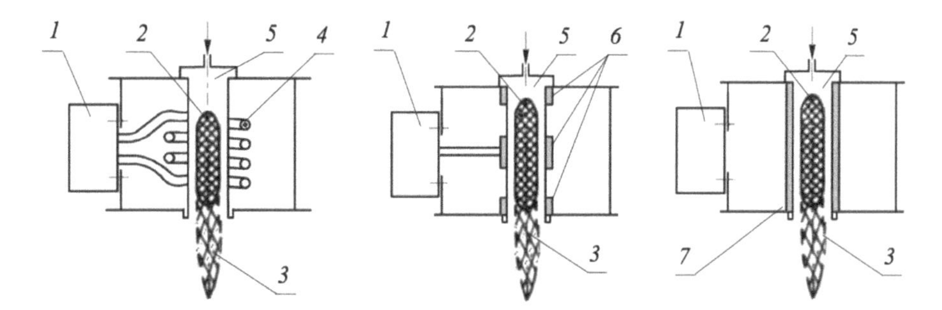

FIGURE 1.15 High-frequency plasmatrons: *a*) High-frequency inductive (HFI) plasmatron; *b*) High-frequency capacitive (HFC) plasmatron; *c*) Ultrashort-wave (USW) plasmatron. 1 – Power supply, 2 – Discharge, 3 – Plasma stream, 4 – Coil, 5 – Discharge chamber, 6 – Electrode, 7 – Waveguide.

to a two-cycle scheme, some with counter flows of plasma, still others are combinations of two HFI-discharges burning in one discharge chamber or combined HFI/HFC plasmatrons supplied with the same two-cycle scheme. *Ultrashort-wave (USW) plasmatrons* operate at frequencies from above 50 to 150 MHz. They are characterized by large volumes of plasmas, ca. 1–2 liters, and low densities of energy, less than 1 W/cm^3. This is achievable for the atmospheric pressure and low specific power (Toumanov, 2003).

In high-pressure ($p > 10$ kPa), direct-current (DC), and radio-frequency inductively coupled plasma (ICP) devices operating with common plasma gases (Ar, N_2, H_2, He) at flow rates of 100–300 lpm, the torch input power can be up to 150 kW. For the same arc current, the anode heat flux for the CO_2/CH_4 plasma is significantly up to three times higher and the overall thermal efficiency of the CO_2/CH_4 mixture is much better (65–75%) than of Ar (<50%), despite the higher heat losses. Thus, at a given current, the CO_2/CH_4 mixture generates a much higher power than the argon mixture does. For example, the torch is capable of operating at a power of up to 70 kW at arc currents of less than 400 A when the plasma-forming gas consists of a mixture containing 70% CO_2 and 30% CH_4. The arc power is limited by the maximum heat flux withstandable by the channel wall, which is ca. $2{\cdot}10^5$ kW/m^2 (Mostaghimi et al., 2017).

The plasma process is governed by a large number of parameters from input power to furnace configuration and can be generated using various types of plasma reactors. Thermal plasma has found its application in coating technologies, synthesis of fine powders, spheroidization with densification of powders, slag metallurgy, a wide range of laboratory and industrial processes, and at various stages of product life cycle until waste destruction and recycling (Samal, 2017). It is especially useful in the case of metal-containing waste and new wastes and has a great potential in recycling. Plasma methods prove to be an efficient and environmentally friendly way for processing of various metal-containing waste types, such as printed circuit boards, red mud, galvanic sludge, zircon, aluminum dross, and incinerated ash. plasma methods, which include DC extended transferred arc plasma reactor, DC non-transferred arc plasma torch, RF thermal plasma reactor, as well as argon and argon–hydrogen plasma jet are particularly useful. In addition, the plasma arc melting technology has a better purification effect on the extraction of useful metals from metal-containing waste, provides a substantial volume reduction of waste materials, and ensures a low leaching toxicity of solid slag. Plasma processing can be applied to all kinds of metal waste materials (Du et al., 2018). Moreover, plasma-assisted pyrolysis and gasification of different sorts of waste can be performed (Sikarwar et al., 2020):

- Using DC plasmatrons: medical waste, wood, municipal solid waste (ca. 2.1 billion tons per year are produced), carpet waste, solid waste, used old tires, polypropylene, agricultural residue, granulated metal powder, C coating from CH_4, vanadium ore
- Using DC-RF hybrid fixed bed reactor: charcoal deformed after treatment
- Using RF plasmatrons: MSW with raw wood, tire powder, rice straw, polyethylene, polypropylene, heavy oils
- Using microwave (MW) plasmatrons: spirulina algae, polyethylene, glycerol

From the industrial perspective, the main advantages of plasma processes compared to other methods are as follows (Vandenabeele and Lucas, 2020):

- Plasma techniques are environmentally clean, require very small quantities of chemicals, and produce little waste.

- Plasma methods involve relatively simple and easily up-scalable apparatus, which enables successive cleaning, activation, and coating deposition without any hazardous handling, simply by changing the processing gas.
- Treatment times (some tens of minutes at most) and numbers of processing steps are also significantly reduced, since components can be exploited directly after the treatment. Thus, plasma processes are economically attractive.
- In the area of nanomaterial plasma surface treatment, especially in nanocrystal plasma synthesis, the resistance of particles to aggregation due to a high degree of surface charging is a very important benefit.

Given the enormous variety of plasma processing and plasma-assisted machining techniques, it seems convenient to follow a classification of methods according to their final effect (Dragobetsky et al., 2012):

1. Material removal methods
   (a) cutting of pre-fabricated material
   (b) shaping of workpiece
2. Joining processes
   (a) plasma welding
   (b) plasma microwelding
3. Additive methods
   (a) plasma cladding
   (b) plasma sputtering

According to the classification provided by Quintino (2014), the third category can be treated as "surface engineering," since this term covers a diversity of technologies that alter the chemistry and properties of a thin surface layer of the substrate, including cladding processes which produce thick coatings. From this point of view, there are several additional methods of surface treatment with plasma aimed at modifying chemical and physical properties of a surface, such as cleaning, activation, etching, coating, plasma ashing, plasma electrolytic oxidation, plasma functionalization, plasma polymerization, and plasma modification of surface layer.

Plasma arc cutting (PAC) is a nontraditional metal removal process which employs a high-temperature and high-velocity constricted arc. The energy content of the plasma is focused through a nozzle with an increased momentum and the intense heat melts and partially vaporizes workpiece material. A power density of 10^8 W/m^2 can be generated, which can vaporize almost all solid materials, allowing to cut stainless steel, manganese steel, high-speed steel, cast iron, and hardened alloys at greater speeds. Moreover, high tolerance PAC effectively cuts titanium sheets and the quality characteristics are better with oxygen as plasma gas than nitrogen because of the oxidation process (Adalarasan et al., 2015). In plasma arc cutting, an electric arc is established between a workpiece (cathode) and a tungsten anode with characteristics of ca. 5,000 A/cm^2, 100 A, and 100 V. The arc causes a localized melting of the workpiece and a gas flux removes the liquid phase of the material. As a rule, PAC employs plasma gases Ar, N, H_2 or their mixtures, the torch is often equipped with a secondary shielding gas flow that surrounds the plasma to confine it and clean the kerf. A variant of the process is carbon arc cutting (CAC), where carbon electrodes and an air flux are applied (Biesuz et al., 2021). Most plasma cutting machines are dedicated to rather simple processing, without loading/unloading automation, and they commonly work with a single piece at a time with limited geometries (Kanyilmaz, 2019). PAC is a productive method for linear cutting or 2D profiles, but if a second cut plan is required, engineers tend to use traditional tools. Nevertheless, plasma arc cutting has acquired a massive ground in the industry (Gani et al., 2021).

Among methods shaping a workpiece, plasma arc drilling is widely applied to materials processing. It is a novel drilling method that has emerged in recent years to overcome such problems as tool wear and low efficiency in the drilling of thick plates or plates made of difficult-to-machine materials. It provides high-quality drilled holes. For drilling of a 12 mm thick mild steel plate, a penetration time of about 0.75 s, a minimum specific energy of about 30 J/mm^3, and a maximum material removal rate of about 300 mm^3/s are reported (Sun and Kusumoto, 2010).

In the *plasma arc welding (PAW) process*, the arc can pass through a nozzle which constricts the arc reducing its cross-sectional area. As a result of increased energy density and velocity of the plasma, the temperature dramatically increases to ca. 25,000°C. Plasma arc welding uses nonconsumable tungsten electrodes and shielding gas (Sahoo and Tripathy, 2021). According to the process current, three types of plasma arc welding processes can be distinguished:

1. Micro-plasma arc welding with a forming current below 15 A. It produces low energy density and low plasma velocity and thus is suitable for thin sheet processing.
2. Melt-in mode plasma arc welding, where current varies between 15 and 400 A. It is usually applied to the welding of thicker, up to 2.4 mm sheets.
3. Keyhole mode plasma arc welding is used for 2.5-mm thick materials. The plasma-forming current works at more than 400 A.

To improve weld efficiency in terms of mechanical strength, weld penetration at a low cost, and faster production, the process can be modified in different ways, e.g., by varying the plasma formation current range, applying different orifice diameters to adjust plasma gas flow rate and thus to increase weld penetration and stability of the arc. Just to name several techniques aimed at increasing the weld efficiency, variable polarity plasma arc welding, double-sided arc welding, ternary gas plasma arc welding, laser-assisted plasma arc welding, PAW–MIG hybrid welding process, soft plasma arc welding, plasma spot welding process, manual pulse keyhole plasma arc welding, or increased power density plasma arc welding can be listed (Sahoo and Tripathy, 2020).

The arc used in PAW has a much higher velocity (300–2,000 m/s) and heat input intensity (10^9–10^{10} W/m^2) than that in conventional gas tungsten arc welding (GTAW). As a consequence, PAW has many advantages over GTAW, namely (Wu et al., 2014):

- PAW offers a greater welding speed, greater energy concentration, and higher efficiency than GTAW, which makes it one of the most effective processes for many applications. PAW gives better penetration than GTAW and welding in the keyhole mode can make full-penetration welds in a relatively thick material with a single pass.
- PAW uses a nonconsumable tungsten or tungsten alloy electrode. The electrode contamination is minimized, which extends its service life. An electrode can usually last for an entire production shift without needing to be reground.
- PAW offers a large tolerance to joint gaps and misalignment. Although the arc is constricted, the plasma column has a significantly larger diameter than the electron or laser beams. As a result, PAW minimizes the need for costly joint preparation and reduces or eliminates the need for filler metal.
- PAW provides a high depth-to-width ratio of a plasma weld compared to a GTAW weld, which can greatly reduce angular distortion and residual stress. Narrower HAZ, fewer internal defects, and better welding processing properties make it an effective method to weld structural components with difficult-to-weld rear sides.

It can be stated that PAW has the potential to replace GTAW in many applications as a primary process for precise joining. However, the PAW equipment, including power supply and electric control systems, is more complex and costly, and the need for water cooling of the torch sets limitations on how small the torch can be (Wu et al., 2016).

Plasma cladding and sputtering belong among additive processing or surface engineering methods. A *plasma cladding* machine can be integrated with three-dimensional CAD software in order to tailor the required pattern. With a fully automated 6-axis CNC device, parts up to 15 m can be cladded to the highest quality, forming wear-resistant coatings and films to improve performance of components. Due to the low heat input, dilution is reduced to a minimum, so that an obtained alloy maintains optimum wear-resistant properties. The heat-affected zone under the surface is very thin, so part distortion is practically absent, which allows processing of thin-walled components. The entire process is highly reproducible; therefore, homogeneous coating deposition can be obtained with a constant and controlled quality (Chandrasekar et al., 2016).

Sputtering is a subset of plasma-material interactions, a phenomena-rich field involving interactions that can occur between the plasma species, including ions, electrons, neutral particles, and radiation on the one hand, and the material surface on the other. Essentially, plasma sputtering belongs to the physical vapor deposition (PVD) category, where vaporized atoms or molecules from a liquid or solid phase are transferred through a vacuum or low-pressure system to be deposited and condensed onto a substrate surface. The ultimate thickness of the obtained thin film varies between 1 and 1,000 nm (Rasouli et al., 2021).

Plasma is also used in other methods of surface engineering, being a valuable tool for obtaining both surface and bulk properties, such as protection against corrosion, oxidation or wear, biocompatibility, wetting, adhesion, durability, catalytic activity, and toughness (Quintino, 2014). In methods based on thermal approach, like *plasma spraying*, fine molten or semi-molten particles are sprayed onto a surface. This technique offers coating thicknesses from several micrometers to over 100 μm using different materials, such as ceramics, plastics, alloys, and composites. Residual stress can, however, negatively affect the stability and, hence, the durability of a coating layer, which can be considered a major disadvantage of the thermal coating methods (Rasouli et al., 2021). In *atmospheric pressure plasma spraying*, powders of sprayed materials are introduced into a plasma torch and melted or partially melted powders at high temperature are accelerated toward the substrate at a high speed. Plasma spraying forms a coating with a lamellae structure and the ability to deposit almost any metals and various combinations of materials. A very high temperature of the arc core allows for use of coating materials with a high melting point such as ceramics, cermets, and refractory materials (Chu et al., 2002). As-obtained plasma-sprayed ZrO_2–8 wt% Y_2O_3 and (Ba,Sr) $Al_2Si_2O_8$ (BSAS)-mullite coatings have been used by NASA as thermal barrier coatings for superalloy components and environmental barrier coatings for SiC/SiC ceramic matrix composite systems, respectively. Yttria stabilized zirconia (YSZ) coating remains stable at high operating temperatures, has a low thermal conductivity of 2 W/(m·K) for bulk and 1.5–2.05 W/(m·K) for plasma-sprayed YSZ. Multilayer ceramic coatings that incorporate mullite, YSZ, and ultra-high-temperature zirconium diboride ZrB_2 are suitable for various systems of thermal protection in rocket exhaust cones, insulating tiles for space shuttles, as well as engine components and ceramic coatings that are embedded into airplane windshield glass (Morks et al., 2013).

Plasma immersion ion implantation (PIII) is also sometimes referred to as plasma implantation, plasma ion implantation, or plasma-based ion implantation. In this process, specimens are surrounded by a high-density plasma and pulse biased to a high negative potential relative to the chamber wall. Ions generated in the overlying plasma are accelerated across the sheath formed around samples and implanted into target surfaces. In PIII, gaseous plasma can be generated by DC or capacitively/inductively coupled radio-frequency glow discharge (rfGD), while metallic plasma can be generated separately or simultaneously by vacuum arc plasma sources. At the same time, many elements can be introduced into the plasma (Chu et al., 2002).

Air or oxygen plasma has found its application in *superior cleaning and activation*, providing advantages over wet chemical and UV/ozone activation in terms of energy input, safety, hazardous waste, corrosion, thermal load, processing time, and versatility in handling a range of material surfaces. In comparison with oxygen or air techniques, plasma activation contains a larger number of variables to be optimized, which further expands the potential for the process improvement (Rasouli et al., 2021).

A cost-effective electrochemical procedure for surface treatment of steels employing plasma is *plasma electrolytic oxidation (PEO)*. It can be mainly applied to valve metals or their alloys including Ti, Al, Mg, and Zr, and recent reports also include non-valve metals like Zn, Hf, Ta, and Nb, as well as diluted alkaline and environmental-friendly solutions containing phosphate, silicate, and aluminate anions. With such wide possibilities, PEO is a promising method able to produce porous, high crystalline, and about 1–100 μm thick oxide coatings. As-obtained coatings have demonstrated good adhesion to the substrate, high surface microhardness, high-temperature shock tolerance, significant resistance to corrosion and wear, and a significant insulation resistance in working conditions (Attarzadeh et al., 2021).

Plasma beam can be used for *polishing* of some metals and alloys. Deng et al. (2018) propose a method of micro-beam plasma polishing using a modified micro-plasma welding machine. Their plasma was operated in a continuous-wave mode, with a maximum electric current of 11 A. The micro-plasma torch consisted of a tungsten electrode bur (d = 1.0 mm), a nozzle (d = 1.2 mm), a plasma protective cap (d = 7 mm), and a pipe liner for centering. Argon was used for the purposes of both ionization and protection of alloy steel from oxidization, at a constant gas flow rate of 6 L/min for protection and a maximum of 0.5 L/min for ionization. The authors concluded that micro-beam plasma polishing could reduce surface altitude differences by 91% and surface roughness by 88%, e.g., from Ra = 4.07 μm to Ra = 0.46 μm. Additionally, micro-beam plasma polishing provided some improvement to the hardness of tested steels and alloys without surface impurity defects.

In *remanufacturing* processes, *micro-plasma transferred arc* (μPTA) is one of the latest options. It can be applied to remanufacturing of die and mold surfaces in a near-millimeter range in order to extend their usage and become a viable economical and cleaner methodology for potential industrial applications (Jhavar et al., 2016). μPTA deposition is advantageous in terms of lower initial setup and running costs with adequate quality of repaired components. A comparison of power consumption per unit weight of filler material and consumption rate of filler material for material deposition using laser between gas metal arc welding and μPTA sources demonstrated that μPTA is a successful alternative for remanufacturing of tool steel surfaces.

Plasma can also be employed in methods of *plasma-enhanced chemical vapor deposition*. The chemical reactions required for the deposition are facilitated by the presence of a plasma arc containing hot electrons. Following deposition of a vapor phase on the substrate, a coating is produced out of metals, ceramics, and carbon-based nanomaterials (Biesuz et al., 2021).

1.12 WATERJET MACHINING

New materials, such as high-temperature alloys, high-performance ceramics, metal-matrix composites, and fiber-reinforced plastics, are examples of materials which have to be machined with adapted conventional machining techniques or by means of new machining processes due to their wear resistance and high strength maintained kept even in aggressive environments or at high temperatures (König et al., 1990).

1.12.1 Waterjet Machining Principle and Influencing Factors

Waterjet machining can be considered a cold cutting process, because it involves the removal of material without releasing heat. This revolutionary technology is able to cut through virtually any material (Yogeswaran and Pitchipoo, 2020). Its application is especially important in the case of polymeric foams, which have been applied for many years, and sandwich structures with aluminum honeycomb cores, which currently experience a significant and growing interest. This type of construction consists of two thin facing layers separated by a core material. With the application of a carbon fiber sheet for sandwich facings, it displays a high stiffness, high tensile strength, low weight, high chemical resistance, high-temperature tolerance, and low thermal expansion. Core shapes and core material can be of various types, including perhaps the most popular honeycomb core that consists of very thin foils in the form of hexagonal cells perpendicular to the facings. The "lightweight design" philosophy is essential to the transportation (automotive, aerospace, shipbuilding) industries; therefore, this type of *new materials* have been adopted and use of sandwich structures has steadily increased in recent years (Yogeswaran and Pitchipoo, 2020). On the other hand, Uhlmann and Männel (2019) demonstrate the feasibility of AWJ in near-net-shape fabrication of three difficult-to-cut materials, namely, titanium aluminide, type Ti-43,5Al-4Nb-1Mo 0,1B (TNM-B1), an MMC composite of a standard titanium alloy Ti6Al4V with 5% titanium carbide (Ti64 + 5%TiC), and zirconium dioxide (ZrO_2).

The first commercial waterjet cutting system was developed in 1971 to cut laminated paper tubes, and in 1980, the process was modified by adding abrasives to the plain waterjet (PWJ). Thereafter, an abrasive waterjet (AWJ) was invented to cut various industrial materials such as steel, glass, and concrete (Liu et al., 2019).

In the waterjet machining process, a material removed from a target workpiece emerges through an erosive venture of abrasive particles traveling at a high velocity. Two primary models of material removal can be pointed out, namely cutting and plowing deformation wear mechanisms, which depend on workpiece material and properties. In this respect, a workpiece can be categorized as ductile, brittle, or composite (Llanto et al., 2021).

During interaction with ductile materials, such as metals, erosion can occur through repeated plastic deformation and cutting action. For brittle materials, the erosion process is predominantly realized through crack propagation and chipping. In the case of composite materials, abrasives penetrate the material and produce breakages that initiate the formation of cracks, which in turn results in delamination. In any case, the erosion mechanism allows for diverse functions of waterjet processing, such as cutting, milling, turning, grinding, drilling, and polishing (Llanto et al., 2021). Abrasive water jet can be successfully used in a remanufacturing process, e.g., to clean components like engine cylinders to be remanufactured (Dong et al., 2014).

Essentially, the abrasive waterjet machining technology uses a jet of high pressure and velocity, water, and abrasive slurry to cut a target material. The process uses a fine-bore nozzle and an orifice of diameter about 0.2–0.3 m to form a coherent, high-velocity jet which has a pressure of up to 400 MPa and a velocity of up to 1,000 m/s. Diameters of the orifice range from 0.08 to 0.8 mm and it is commonly made of sapphire, ruby, or diamond (Saravanan et al., 2020). Figure 1.16 presents cutting heads for plain or pure waterjet and abrasive waterjet machining. It should be emphasized

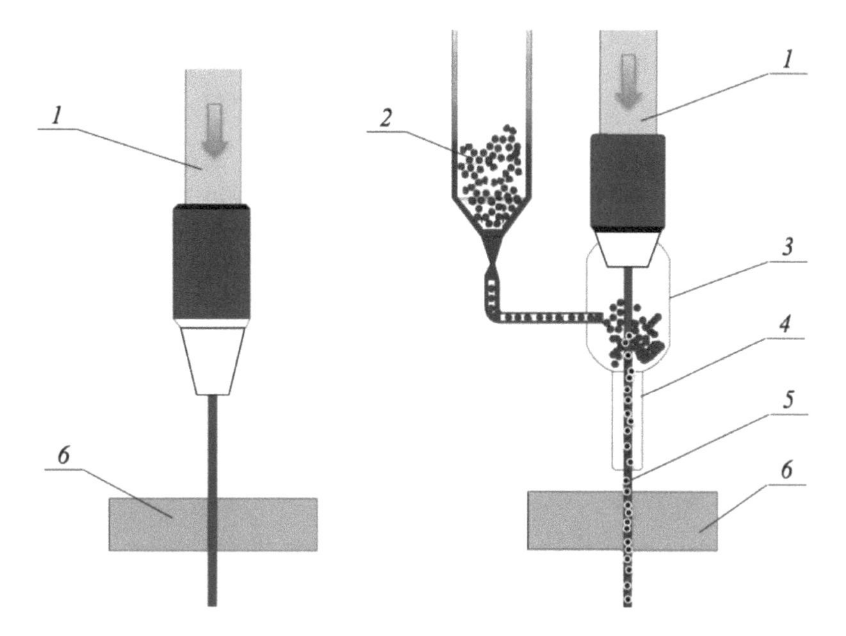

FIGURE 1.16 Cutting heads: (a) Plain/pure waterjet (PWJ); (b) Abrasive waterjet (AWJ). 1 – Water under pressure, 2 – Abrasive hopper, 3 – Mixing chamber, 4 – Focusing tube, 5 – Water and abrasive slurry, 6 – Workpiece.

that both techniques still find their applications in the cutting of different materials. Namely, PWJ is commonly employed for machining weak materials such as foams and polymers or when foreign material contamination has to be avoided, but for cutting high-strength materials such as metal alloys or ceramics, the use of AWJ is necessary (Bergs et al., 2020).

Abrasives are categorized as natural, zirconia alumina, glass, steel, and copper, with different characteristics such as level of hardness and grit shape. For example, those widely applied in AWJ glass beads have mineral hardness MOHS 5-6, in garnet and ceramic beads, MOHS 7-8, black corundum, black and green silicon carbides, as well as white and brown fused alumina MOHS 9, while steel grit and shot, stainless steel cut wire, and copper slag exhibit hardness of 40–60 HRC (Llanto et al., 2021). Natarajan et al. (2020) claimed that more than 80% of industries consume abrasives of the garnet type only.

A harder abrasive demonstrates a greater MRR and depth of cut, providing higher machining efficiency but potentially cutting the lifespan of focusing tubes. On the other hand, a harder workpiece requires a harder abrasive particle. Thus, selection of an appropriate abrasive size and type depends on the hardness of a workpiece. The range of abrasive sizes is universally indicated by mesh size (#), as follows:

- mesh #40–60 of sizes 250–400 μm
- mesh #80–100 of sizes 180–210 μm

- mesh #120–150 of sizes 90–105 μm
- mesh #180–220 of sizes 70–88 μm
- mesh #240 and higher of sizes below 60 μm (Llanto et al., 2021)

Moreover, it is possible to introduce a freezing agent into the water jet to produce ice grains in the stream and thus to increase MRR (Korzhov, 2006). The author emphasizes additional benefits of this solution, such as minimized distortion of stream characteristics, or increased wear resistance of nozzles due to the protective function of the ice layer which appears on their working surface.

Depending on whether a workpiece is completely penetrated or an individual cut into a defined depth is produced, waterjet machining can be subdivided into two categories (Bergs et al., 2020):

1. Cutting through (CT)
2. Controlled depth machining (CDM)

CT is the standard industrial waterjet application. Here, the jet strikes a workpiece surface and its erosive force removes the material, cutting a narrow groove into the workpiece material. When the jet does not cut completely through a workpiece, then CDM takes place. With an increased process complexity, CDM AWJ offers the possibility of shape machining with the potential to substitute conventional milling processes in some applications (Bergs et al., 2020).

The main influencing factors of the AWJ machining process are as follows (Natarajan et al., 2020):

- *Waterjet pressure* is an important process parameter, since kinetic energy of AWJ depends on the pressure level of water. When a certain threshold pressure is not reached, no material removal takes place. On the other hand, pressure equal to the critical value represents a limitation to effective cutting. In the working range, waterjet pressure is directly proportional to penetration depth and material removal rate.
- *Traverse rate* indicated by mm^3/min determines the quality of cut surfaces; its major influence is determined by exposure time. A lower traverse rate enhances surface quality as more abrasive particles are able to impinge on the workpiece surface. It also affects the cutting rate of the process.
- *Abrasives* of various natural (garnet) and artificial (silicon carbide, aluminum oxide) types are used. Abrasive particle size, shape, and hardness have a significant influence on AWJ cutting performance so that a greater hardness of work material requires harder application of harder abrasives. An increased size of abrasive particles increases particle disintegration. A lower depth of penetration and material removal rate can be consequences of a higher limit of abrasive particle size as impingement frequency on the target material surface is reduced.
- *Abrasive mass flow rate* has an influence on AWJ material removal rate. An optimum supply of abrasives yields a higher cutting performance with

a better surface finish and depends on the diameter of the focusing nozzle. The rate of abrasive mass flow is usually expressed as kg/min.

- *Standoff distance*, i.e., the distance between a target material and the nozzle, is usually maintained by an optimal level in mm.
- *Jet impingement angle* is associated with tilting of the cutting head and is defined as the angle between the initial AWJ flowing direction and the target material surface. Two different types of jet impingement angles used in the machining process are distinguished, namely forward and backward. Any change in the jet impingement angle eventually changes the jet attack angle on the target material with an effect on the mode of erosion. Use of a changeover jet impingement angle has a great impact on AWJ cutting performance without any additional costs.
- *Work material* in the AWJ machining process can have any size and shape. The major phenomena found in the work material after the machining process are abrasive contamination and striation formation. The former for soft materials is higher than for hard materials, while the latter is significantly present in a hard material cut surface rather than in soft materials. This happens due to an increased abrasive attack angle at a larger cutting depth.

The striations are curved opposite to the cutting direction and their curvature depends on workpiece material type and thickness, AWJ intensity, and cutting velocity (Orbanic and Junkar, 2008). While the top part of the surface is quite smooth, the bottom part is rough and wavy. A striated surface is characterized by different parameters, typically lag of stria, curvatures of stria, and changing surface roughness or waviness related to the above parameters.

It should be added that effective cutting angles φ depend on the type of workpiece material (Averin, 2017). The author claims that the most effective destruction of a brittle material occurs when the cutting angle is 90°, whereas for ductile materials, the optimum cutting angle is about 20°. A new experimentally established coefficient, k_φ, with a value between 0 and 1, expresses the influence of angle φ of waterjet attack on cutting efficiency. Its calculation is different for the two types of material.

For brittle materials:

$$k_\varphi = 0.99\exp\left(-0.5\left|\frac{\varphi - 90}{28.4}\right|^{1.77}\right) \tag{1.12.1}$$

For ductile materials:

$$k_\varphi = 8437\sin\left(\frac{\varphi}{68049}\right)\exp\left(\frac{-\varphi}{20.5}\right) \tag{1.12.2}$$

These equations obtain for values from 0° to 180° (1.12.1) and for values from 0° to 90° (1.12.2), respectively (Averin, 2017).

1.12.2 Main Components of a Waterjet Machining System

A waterjet machining system essentially consists of a high-pressure pump, cutting head, cutting table with a coordinate motion system, abrasive hooper, and flow control system (in the case of AWJ), and a computer-based controller (Korzhov, 2006). In addition, the system may be equipped with collision prediction and resolution systems, a system of several cutting heads, a mechanical system of initial pre-machining drilling, or a system that disseminates energy of abrasive water jet after machining the workpiece and collects the used abrasive, etc. AWJ systems may be automated to various degrees, including robotic waterjet machining systems.

A *high-pressure pump* is the core element of a waterjet cutting system that drives pressurized water in the nozzle. As a rule, it employs a special double-acting or single-acting multiplier, but its particular design depends on machining conditions, such as pressure drop or a required fluid flow (Korzhov, 2006). This way, expected efficiency and quality performance can be achieved.

The *cutting head* plays a role in forming a final high-pressure jet with characteristics appropriate for cutting. Essentially, the cutting head consists of an orifice, a mixing chamber, and a focusing tube, as depicted in Figure 1.17.

Abrasive particles are fed by air into the mixing chamber, whereby the resulting abrasive jet travels through the focusing tube and momentum is transferred to the particles. When the accelerated abrasive jet exits the focusing tube, it impinges on a workpiece causing the removal of its material. It is common practice to maintain the standoff distance, i.e., the distance between the tip of the focusing tube and the workpiece, between 1 and 2 mm, which is widely accepted as the optimal value for most scenarios (Copertaro et al., 2020). Improper alignment of the orifice in the cutting head can cause additional wear in the mixing tube walls shortening its life and reduce the efficiency of AWJ. A well-aligned and coherent AWJ stream ensures the most efficient cutting, enhances the cutting force, and reduces the kerf taper in the cut profile. Moreover, a greater orifice diameter reduces cutting efficiency due to a higher flow rate of water that lowers the concentration of abrasive particles (Natajaran et al., 2020). As far as nozzle inner surface is concerned, it was demonstrated that maximum efficiency can be reached using a catenoid profile, while a conoid profile is better than conical (Gusev et al., 2012).

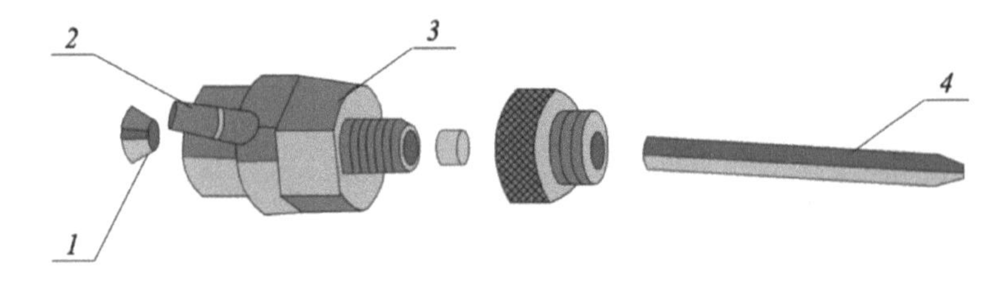

FIGURE 1.17 Example of a waterjet cutting head: 1 – Orifice, 2 – Abrasive inlet, 3 – Mixing chamber, 4 – Focusing tube.

Among the various classifications of waterjet cutting heads, the following can be recommended (Korzhov, 2006):

1. Dynamically improved cutting heads equipped with special structural elements
2. Abrasive waterjet cutting heads, where the most feasible designs are able to introduce the abrasive slurry into the water jet with a minimal distortion of its hydrodynamic characteristics
3. Combined nozzle heads, such as dual-head or double-head systems, where up to four cutting heads can work simultaneously

According to the system configuration, jet tools can also be categorized by their phase composition (Bergs et al., 2020):

1. Standard PWJ (one-phase – water)
2. Injection PWJ (two-phase – water, air)
3. Suspension AWJ (two-phase – water, abrasive)
4. Injection AWJ (three-phase – water, abrasive, air)

High-pressure distribution system. Waterjet cutting systems work with a very high water pressure of 400 MPa. The jet leaves the cutting nozzle at 1,000 m/s, which is three times the speed of sound. The pump must be protected from cavitation and the pressure must be delivered from the pump to the cutting head. Fixed and movable pipelines are applied together with special high-pressure hinges or specially shaped spiral tubes.

Abrasive hooper. Korzhov (2006) specifies two main systems delivering abrasive powder to the mixing chamber of the cutting head, namely based on vacuum or air pressure. The vacuum system works as a pulverizer, while in the latter the powder is forced into the mixing chamber by pressurized air.

1.12.3 Main Advantages of AWJ Machining

Lack of thermal distortion, high machining versatility, high flexibility, and small cutting forces are among the most distinct advantages of AWJ machining. No electrical or thermal energy is used, therefore, many material defects can be ignored (Saravanan et al., 2020). Korzhov (2006) emphasizes the following benefits:

- *Elimination of thermal impact* due to a continuous cooling effect of the water jet. No significant rise in workpiece temperature is noted, which is a decisive factor in the machining of thermosensitive materials. Cutting forces of between 1 and 100 N and temperatures between +60 and +90°C in the cutting area eliminate any deformations of the workpiece. It should be noted that no other technique can be performed without thermal effects on metals.

- *Universality of the process* means AWJ machining applicability to virtually any material. In the case of composites or multicomponent sandwich-like structures, the water jet does not cause disruption of the original structure and thus initial properties of a material are retained after AWJ machining.
- *Ability to reproduce complex profiles and shapes* at any declination angle. In this respect, the water jet can be compared with a point-cutting tool with all its benefits. Especially while machining a brittle material like glass, AWJ allows for producing shapes unachievable by other techniques. While straight cutting of glass with diamond is more efficient, no other cutting technology can produce such a complex shape out of glass.
- *Good surface quality* of roughness *Ra* between 0.5 and 1.5 μm is achieved, so that additional finishing is not necessary in many cases.
- *Technological flexibility* of the process is due to the unique properties of the cutting tool. It is not worn, so there is no need to replace or sharpen it, tool load on the workpiece is minimal, backward pressure is absent, since there is no direct contact between the cutting head and the workpiece. The same tool can perform different technological operations and its low tangential force allows for avoiding fixation of the workpiece. It is possible to use the waterjet cutting head even 200 m away from the pump, and one pump can feed two or more cutting heads operating either on the same table or on different tables. AWJ machining can be performed in various conditions, e.g., at a height of hundreds of meters and even under water.
- *Economic benefits* can be derived from the high cutting speed. The operation can be started at any point on the workpiece surface with no need for pre-drilling. The small width of a slot after cutting indicates savings of the cut material. Added to all that, water consumption is rather small despite high pressure and is kept in the range of 3–4 l/min.
- *Automation of the process* is very easy, digital control can be applied together with optical inspection devices, so that AWJ can be realized by 6-axis full robots.
- *Availability* of the process is easy due to the simple and relatively cheap media like water and quartz sand.
- *Safety* conditions are assured because no heat is accumulated during AWJ machining, with no danger of fire or explosion. There is no radiation, dust, or smoke either, and the noise level is between 85 and 95 dB.

1.12.4 Limitations and Environmental Impact of AWJ Machining

Though well established, the AWJ machining process has some limitations that prevent the wider industrial application of this method. These include generation of a higher volume of secondary wastage following machining and of heat developed in the primary impact zone, abrasive contamination, taper and striation formation, rough surface, and low energy-transfer efficiency from the nozzle to the workpiece, which causes a low depth of penetration and low material removal rate (Natarajan et al., 2020). Additional limitations are listed by Korzhov (2006):

- Structural difficulties with a high-pressure water supply and cutting heads
- Relatively low wear resistance of the nozzles
- Relatively high energy consumption
- Difference between nominal and real characteristics which may prevent successful cutting of some materials
- In some small factories, relatively expensive waterjet cutting equipment may appear unprofitable

Environmental impact of AWJ machining may be analyzed from two perspectives, namely its application to 6R processes (reduction, repair, reuse, recover, remanufacturing, recycling) and the environmental friendliness of AWJ itself. In this respect, the abrasive waterjet cutting technology provides some benefits from both perspectives.

Apart from remanufacturing processes, AWJ can be utilized *to crush waste printed circuit boards* for the first time. A recent report (Yang et al., 2021) demonstrated that AWJ cutting was capable of breaking ca. 14-mm thick waste universal circuit boards into small particles below 1 mm. As a result of one-step processing, metals and nonmetals were well dissociated. All the cutting debris of the central processing unit, mainly Cu and Ti particles, were smaller than 150 μm, and metals were completely dissociated. The results implied that AWJ cutting has a huge potential for enhancement of the material removal indicator and can be a promising environmental-friendly method for recovering metal resources from e-waste.

On the other hand, as new high-tech abrasives (such as ceramic abrasive, silicon carbide abrasive) continue appearing, the proportion of abrasives cost in wet abrasive blasting cleaning is increasing. Thus, *reuse of the abrasives* becomes more and more important. To address the issue, a recycling system was proposed by Dong et al. (2014). It consisted of four parts: abrasive water jet cleaning, abrasive collecting, waste abrasive separating, and sewage treatment system. The authors demonstrated that the recycling of abrasives reduced the cost of AWJ cleaning and environment pollution, making the method more environmentally friendly and more economical. Schramm et al. (2020) reported the recycling potential of suspension fine jet processing estimated by an integrated technical–economic evaluation. The authors showed a general particle size reduction, so that without cut material 72%, with ceramic material 83–87%, and with aluminum 80–81% of the abrasive material can be reused. Costs of recycling and those of single use of abrasives with subsequent disposal were analyzed and it proved that recycling abrasives can reduce the costs by 61.27% compared to a single use of abrasives. In addition, the authors identified demand for new abrasives as the main variable influencing the result.

1.13 TRENDS IN JOINING TECHNOLOGIES

Böllinghaus et al. (2009) emphasize that joining technology today derived high-technological procedures from conventional process technologies and is rapidly developing toward microjoining and even nanotechnology. They suggest that increasing attention is paid to optimized material utilization considering local work conditions of components. For instance, "tailored blanks" denote sheet products characterized by a local variation of sheet thickness, sheet material, coating, or material properties. The concept of *tailored blanks* can be divided into four subgroups, namely tailor welded blanks (joining materials with different grades, thicknesses, or coatings by a welding process), patchwork blanks (locally reinforcing a blank by adding another), tailor rolled blanks (continuous variation of sheet thickness achieved by a rolling process), and tailor heat-treated blanks (material properties adapted by a local heat treatment) (Merklein et al., 2014). Furthermore, combinations of steel with light metals or plastics are increasingly applied to lightweight constructions (Böllinghaus et al., 2009).

1.13.1 Remarks on General Trends

According to Middeldorf and von Hofe (2008), it is crucial to take the following issues into consideration in technical design: availability of a large and increasing variety of structural materials, increasing requirements of component remanufacturability, and effective recycling of EoL products. In this respect, they formulated several important remarks:

- Nowadays, both investment goods such as building structures, all types of vehicles, installations, and machines, and everyday consumer goods like furniture, household appliances, and electronic devices consist of a multitude of various materials. These materials joined together provide combinations of functional characteristics and resource savings. Thus, different materials are gathered not only in a device or another product but also in smaller units or even single components. This is possible only on the condition that a joint meets all functional requirements, such as strength, ductility, hardness, corrosion resistance, etc.
- Rapid increase of structural materials with enhanced properties, and especially new materials with special properties, generates demand for appropriate joining technologies. For example, a combination of textile fibers with concrete, or nanoparticles with plastics, provides qualitatively new materials which require a completely new joining technology to be integrated into a component.
- Computational technologies make it possible to predict functional characteristics of parts or units made of new materials, thereby rapidly expanding areas of their implementation.
- When new materials and joining processes are designed, it is crucial to consider the possibility of their recovery in the early phase of exploitation. On the one hand, a damaged component should be recycled, and on the other, reuse and remanufacturing may return it to a working engineering system.
- Perhaps today it appears to be economically profitable to replace a damaged component with a new one, yet from a wider perspective this contradicts any resource-saving strategy. In the context of transition to circular economy, recovery questions will arise again and again.

1.13.2 Promising Joining Technologies

Böllinghaus et al. (2009) wrote that gas-shielded arc welding retained a dominant position, but its development slowed down. Resistance welding was expected to prevail due to its process performance and productivity, but in some applications it would be replaced by mechanical joining and adhesive bonding, or hybrid processes such as a combination of resistance spot welding with adhesive bonding. Due to the rapid development of laser technology and its transfer to applications, its continuing growth is expected, while the rise of electron beam welding will possibly slow down.

The *electron beam welding* process is an advanced fusion welding technique able to fabricate structural parts with a high dimensional precision inducing minimum thermal stresses and distortion (Chowdhury et al., 2021). Although electron beam welding has been known for over 60 years, in the last decade it has attracted the growing interest in science and industry. In comparison with other methods of joining, electron beam welding has the following advantages (Węglowski et al., 2016):

- It has an extremely high power density of about 107 W cm–2 at the beam focus.
- Energy transfer occurs by the conduction of heat across the surface of a workpiece itself.
- As no edge preparation is necessary, regardless of workpiece thickness, the filler metal is generally not required.
- High welding speed results in narrow welds and heat-affected zones with little distortion of the workpiece.
- Inertia-free oscillation of the electron beam makes it possible in many cases to join materials otherwise considered unsuitable for welding.
- Variable working distance allows workpieces of widely differing shapes to be welded.
- As welding is carried out under a vacuum, no consumables (gases, fluxes) are required to protect the weld pool from oxidation.
- Short evacuating times can be achieved by adopting a working chamber to suit the number and size of workpieces.
- Computer monitoring and control of electrical and mechanical welding parameters are possible.
- Welding parameters, and thus the quality of the welds produced, are highly reproducible and consistent.
- At accelerating voltages above 60 kV, lead shielding is directly bonded to the welding machine in order to prevent X-ray emission.
- Beam powers of far less than 1 to 300 kW are available for welding material thicknesses of less than 0.5 to 300 mm.
- Machines are available for welding variable one-off components as well as for use in mass production operations like in the automotive industry.
- Simple longitudinal weld seams can be made as well as complicated three-dimensional components requiring the use of programmed welding parameters and workpiece manipulation.

- Structural steels, alloy steels, non-ferrous metals, and even gas-sensitive special metals can be successfully welded.

Since vacuum limitation is usually mentioned among the main drawbacks of electron beam welding, it should be noted that processes without vacuum are now available and gaining new industrial applications (Middeldorf and von Hofe, 2008).

Laser welding has gained great popularity as a promising joining technology with high quality, high precision, high performance, high speed, good flexibility, and low deformation or distortion, and the number of its applications is increasing due to ease of use with robots, reduced man-power, full automation, systematization, production lines, etc. (Katayama, 2018). Its main advantage is high energy density and hence the ability to melt the area close to joint edges without affecting a large area of the workpiece. Other benefits are high welding speed, achievability of a narrow and deep weld, good mechanical properties, and low structural distortion (Xiao and Zhang, 2014).

There are also advantages of laser compared to conventional welding in respect of joining of dissimilar materials (Martinsen et al., 2015). Laser welding is able to fuse dissimilar metals and alloys and overcome the problems with large differences in thermal conductivity. The small size and accuracy of the beam provide control of the microstructure, and a small heat-affected zone results in rapid solidification and less grain growth. Using through-transmission laser welding (TTLW), where the upper part is transparent for the laser beam and the lower part absorbs the laser energy, thermoplastic polymers can be welded, as well as dissimilar polymers, and even polymer–metal bonds are possible.

Friction stir welding (FSW) is a solid-state process where a rotating stirring tool softens a material with frictional heat and welded materials are mechanically stirred and bonded (Martinsen et al., 2015). The process minimizes brittle intermetallic phases, metallurgical incompatibility, differences in melting points, and thermal mismatch. The friction stir process family includes friction stir processing, friction stir casting, friction stir micro forming, friction stir powder processing, friction stir channeling, and friction stir spot welding. Among the advantages of FSW are its environmental friendliness, application of non-consumable tool, requirement of little or no post-processing, removal of residual stresses, corrosion resistance; the process doesn't require the addition of a filler material or oxide removal, cost effectiveness, and the possibility to create dissimilar material joints. Middeldorf and von Hofe (2008) emphasize that due to low temperatures, significantly below the melting point, only minor metallurgical alterations take place in joint materials. However, large forces applied to joint details require additional strong equipment, which may be considered a disadvantage of FSW.

Structural *adhesive bonding* is a joining technology with many advantages, especially in aerospace applications in the context of *structural repair*, a major challenge with advanced composites because of their inherent complex damage behavior (Katnam et al., 2013). In their review, Martinsen et al. (2015) listed several theoretical mechanisms of adhesion:

1. Adsorption theory
2. Chemical bonding
3. Diffusion theory
4. Electrostatic attraction
5. Mechanical interlocking
6. Weak boundary layer theory

However, the authors added that none of these theories were capable of providing a comprehensive explanation for all types of adhesive interaction and joints. They also listed several factors defining adhesive bond performance, such as lifetime or robustness (Martinsen et al., 2015):

- Physical and chemical properties of adhesives
- Nature of adherents whose materials are to be joined
- Type of bonding surface pre-treatment (preparation), very important for joint quality
- Surface wettability
- Joint design, i.e., appropriate adhesive choice, which plays a crucial role in the selection of an adhesive and curing method

Katnam et al. (2013) claim that the resulting bond strength must be sufficient to ensure that stresses are safely transferred between the two adherents providing a shear dominant stress state and inducing minimum peel (i.e., through-thickness normal) stresses in the adhesive layer. For bonded structural composite repair, single- or double-sided doubler patches, scarf or stepped scarf patches can be bonded to a damaged structure. In general, bonded scarf patches can be considered the most efficient joints for bonded repair, since they have no eccentricity of the load and produce minimal peel stresses (Katnam et al., 2013).

Resistance spot welding (RSW) has been the mainstream solution for steel-dominated automotive bodies, but today it is applied to obtain dissimilar material joints such as Al–steel, Mg–steel, glass–steel, Cu–Ni-plated steel, stainless steel–Ti, Ti–Al, and Al–Mg combinations (Martinsen et al., 2015). The authors list among benefits of the method its flexibility and easy automation, and the fact that since the weld gun electrodes are clamped, small gaps between the sheets can be leveled.

Hybrid joining techniques as a combination of adhesive bonding with a spot-joint, such as a resistance spot-weld or a clinched joint, are widely used in car body manufacturing (Meschut et al., 2014). Application of heat-curing epoxy-based structural adhesives significantly increases joint strength and the adhesive layer also isolates and seals the joint area. The latter is a very important aspect for multi-material design, where a difference in electrochemical potential of joined materials can result in accelerated corrosion effects. Today, mechanical joining technologies like clinching and self-pierce riveting are often combined with adhesive bonding. Trying to summarize the advances in hybrid joining technologies, the authors emphasized that classic mechanical joining combined with local conditioning as well as some new approaches show sufficient load-bearing capabilities for automotive applications. All of these methods have individual advantages and disadvantages, and a decision to apply one or more technologies in a particular production depends on requirements like production volume, required strength, material flexibility, costs, etc. (Meschut et al., 2014).

Advancing miniaturization of devices brings challenges to microscale joining technologies. Essentially, effective *microjoining* has been an integral part of manufacturing for many decades in microelectronics, medical, aerospace, and defense industries (Zhou and Hu, 2011). Microjoining processes can be grouped like traditional methods, in respect of solid-state bonding, fusion welding, brazing/soldering, and adhesive bonding. Thus, Jain et al. (2013) described microwelding, fusion microwelding, electron beam microwelding, laser beam microwelding, resistance microwelding, solid-state microwelding, ultrasonic microwelding, and microjoining by adhesive bonding. However, as Zhou and Hu (2011) underlined, further down the road of miniaturization to nanoscale, there is an emerging need to join nanoscale building blocks, such as nanowires and nanotubes, to form nanoscale devices and systems, e.g., micro-electro-mechanical systems (MEMS) and nano-electro-mechanical systems (NEMS), and then to join these to the surroundings integrating them into micro- and macroscale devices and systems. The authors emphasized that the term "microjoining" is already popular in industries, while "nanojoining" is relatively new and often used in relation to conventional macroscale welding and joining. It should be noted, however, that various soldering processes and alloys, widely used in microelectronics interconnection and packaging, have been modified and further developed for the purpose of *nanojoining*. Focused electron beam, widely used in micro- and macrojoining, is also applied to join metallic nanomaterials. Other examples named by Zhou and Hu (2011) include ultrasonic welding, cold welding, laser welding, resistance welding, etc. The authors also point out the main challenges of micro- and especially nanojoining technologies, such as surface roughness and contamination, which significantly increase surface areas and reduce volumes. These may cause difficulties in the manipulation of nanoscale building blocks and damage to their internal structure and properties.

1.14 METHODS INCREASING MACHINING ACCURACY

To ensure machining accuracy, a scientific approach is required. It involves the process of identifying and analyzing potential challenges, analysis and deeper understanding of impacts, as well as the development of methods to minimize accuracy issues (Brecher et al., 2016). For example, it was found that limitation of dynamic properties for CNC machine tools leads to the contouring error and becomes a vital issue that may decrease the machining accuracy of high-performance parts with complex curved surfaces (Jia et al., 2018). An extensive review of ultrafast laser micromachining accuracy was published by Cheng et al. (2013). It discussed effects of some very important parameters such as laser pulse duration, plasma and polarization, as well as some techniques employed for the increase of machining accuracy and throughput. It was shown that the shorter the laser pulse duration, the shorter the heat diffusion depth, thus improving machining accuracy. On the other hand, nonlinear light self-focusing and its interaction with the ambient air may seriously affect machining accuracy. Polarization control and helical drilling improved machining quality and, especially when combined, provided a higher machining accuracy. Chavoshi and Luo (2015) reviewed machining accuracy for hybrid micromachining processes aimed to improve the machinability of hard-to-machine materials, tool life, geometrical accuracy, surface integrity, and machining efficiency. Analysis of accuracy issues of electrochemical nanomachining can be found in Han et al. (2020). In direct-writing electrochemical nanomachining, machining accuracy depends on the size and shape of a tip orifice, as well as the hydrophilicity of both the tip and substrate. For template electroforming in the scale of a few nanometers, the key issue is to avoid defects and cavities in the electroplating layer, especially for fabrication of 3D nanostructures with high aspect ratios.

In general terms, the quality coefficient B_{Ki} can be introduced to assess realization of i-th quality parameter. Conditions for successful implementation of various methods aimed to improve machining accuracy can be written as follows:

$$B_{Ki} = 1 - \frac{\left| \left(\frac{\sum_{j-1}^{J} \alpha_{ji} \left|\alpha_j - \alpha_j^H\right|}{T_j} \right) + \alpha_i - \alpha_i^H \right|}{T_i + \frac{\sum_{k-1}^{K} \left|\Delta\alpha_{ki}\right|}{T_i}} \geq B_{Ki}^T = 1 - \frac{\left|\alpha_i^T - \alpha_i^H\right|}{T_i}, \qquad (1.14.1)$$

$$B_{Ki} = 1 - \frac{\left| \left(\frac{\sum_{j=1}^{J} \alpha_{ji} \left|\alpha_j - \alpha_j^N\right|}{T_j} \right) + \alpha_i - \alpha_i^N \right|}{T_i + \frac{\sum_{k=1}^{K} \left|'' \alpha_{ki}\right|}{T_i}} \geq B_{Ki}^R = 1 - \frac{\left|\alpha_i^R - \alpha_i^N\right|}{T_i}$$

where $B^R{}_{Ki}$ denotes required value of the i-th quality parameter,

α_i, α_j – values of i-th and j-th quality parameters obtained during precision machining and in the previous operations, respectively,

$\alpha^N{}_i$, $\alpha^N{}_j$ – nominal values of i-th and j-th quality parameters obtained during precision machining and in the previous operations, respectively,

$\Delta\alpha_{ki}$ – improvement of the i-th quality parameter obtained through k-th action dealing with a machining accuracy issue,

T_i, T_j – acceptable values of i-th and j-th quality parameters obtained during precision machining and in previous operations, respectively,

α_{ji} – factor expressing impact of j-th on i-th quality parameter,
α^T_i – normalized value of i-th quality parameter.

Parameters α_i, α_j, α_{ji}, and $\Delta\alpha_{ki}$ should be initially determined for concrete machining conditions based on technical characteristics of technological equipment and theoretical relations, and then experimentally evaluated and specified. Normalized deviations from each i-th quality parameter will depend on narrowing acceptable tolerances which ensure the technological accuracy of machining.

For instance, the machining accuracy of drilled holes can be assessed with various quality parameters such as diameter tolerances, roundness, cylindricity, axis linearity, perpendicularity, coaxiality, runout, or positioning tolerances. The surface quality of hole may be assessed using 2D or 3D surface geometry parameters, microhardness, or deformed layer depth and intensity. Recently, surface integrity parameters have gained importance because of their correlation with functional performance. Among them are surface and sub-surface characteristics, such as residual stresses or surface flaws/anomalies/imperfections (Jawahir et al., 2011).

It should be noted that the implementation of a particular tool material with respective wear compensation mechanisms is dependent on machining conditions and requirements of tool dimensional stability. Thus, multi-tool machining may introduce substantial differences to the dimensional stability of particular tools, reducing the overall accuracy performance and efficiency. In that case, the required quality may be achieved by means of appropriate work intervals between dimensional tool calibrations. In hybrid or simultaneous multi-tool machining, accuracy may be ensured using the so-called *load factor* that characterizes relative dimensional durability of different tools of a multi-spindle machine or different working edges of a combined tool. Respective load factors denoted as K_{LMi} for a single tool in a multi-tool machine and K_{LEi} for a single cutting edge of a combined tool can be calculated as follows:

$$K_{LMi} = \frac{L_{Mi}}{L_{Mmax}}, \tag{1.14.2}$$

$$K_{LEi} = \frac{L_{Ei}}{L_{Emax}}, \tag{1.14.3}$$

where L_{Mi}, L_{Mmax} denote dimensional durability (10^3 m) of i-th cutting tool, including combined ones, and maximal dimensional durability of a machine tool, respectively, L_{Ei}, L_{Emax} – dimensional durability (10^3 m) of i-th cutting edge and maximal dimensional durability of a combined cutting tool, respectively.

The dimensional durability of a tool can be defined using the average length of its working path before an acceptable wear level is reached. It can be calculated from the following formula:

$$L_T = 10^3 \cdot \frac{T_w}{j_w}, \tag{1.14.4}$$

where T_w – acceptable dimensional wear of a tool (mm), and j_w – dimensional wear rate (μm/10³ m).

The dimensional durability of cutting edges of a combined tool is dependent on the compensation mechanisms applied during machining. In this respect, it may be classified as follows: L_{E0} – without compensation, L_{Ev} – with a vibrational protective element, L_{Ep} – with partial compensation, and L_{Ef} – for the machining with full continuous and periodical self-compensation of edge dimensional wear. The respective equations for these parameters are listed below:

$$L_{E0} = 10^3 \cdot \frac{(T_{wc} + T_{ws})}{(j_{wc} + j_{ws})}, \tag{1.14.5}$$

$$L_{Ev} = 0.5 \cdot 10^3 \cdot \frac{T_{wc}}{j_{wc}}, \tag{1.14.6}$$

$$L_{Ep} = 10^3 \cdot \frac{(T_{wc} + T_{ws})}{(0.5 j_{wc} + j_{ws})}, \tag{1.14.7}$$

$$L_{Ef} = 0.5 \cdot 10^3 \cdot \frac{T_{ws}}{j_{ws}}, \tag{1.14.8}$$

where T_{wc}, T_{ws} – dimensional wear (mm) of cutting and side guiding elements, respectively, acceptable for required accuracy and surface quality;

j_{wc}, j_{ws} – average values of radial wear rate (μm/10³ m) of cutting and side guiding elements, respectively.

Industrial practice indicates that in mass production, the load factors K_{LMi} and K_{LEi} should not fall below 0.7 and 0.85, respectively. Calculations based on (1.14.2)–(1.14.8) can provide powerful support at the design stage for an appropriate choice of tool materials, considering machining conditions and especially methods of wear compensation. In this respect, an adaptive tool wear compensation strategy is noteworthy, where the traditional linear compensation method is combined with real-time tool wear sensing (Belotti et al., 2021).

It should be emphasized that as mechanical characteristics of machined materials are enhanced, the efficiency of some methods aimed to ensure machining accuracy increases, too. Namely, since some tool wear compensation methods have nearly no effect on the machining accuracy of non-ferrous metals, their application in this area is negligible. These methods, however, are applied in the case of cast iron and are more effective when steels are machined.

2 Contemporary Methods of Protection and Restoration of Components

From the very moment a material is released from the point of production, it is subjected to some form of material degradation and no known service environment provides perfect immunity to this process (Batchelor et al., 2011). Almost any known natural phenomenon contributes to damage mechanisms that cause premature failure of components and devices. Forms of materials degradation can be classified as physical, chemical, or biological phenomena which usually occur in combinations, as illustrated in Figure 2.1.

Batchelor et al. (2011) emphasize that normally multi-phenomena modes of damage take place and protective measures are ineffective when a single-phenomenon mode of damage is mistakenly assumed. Among examples of interactions between degradation processes, the authors named corrosive-wear, which is mechanical wear accelerated by chemical damage to a worn material (physicochemical phenomena). Biochemical phenomena occur, e.g., when not only organisms themselves eat artificial materials, but also waste products of bacteria are destructive to materials. If a metal component is implanted in the human body, it is a subject of biological, chemical, and physical phenomena imposing severe stresses and accelerated materials degradation.

Under the influence of various degradation factors, an adequately protected material can retain its characteristics longer, as shown in Figure 2.2. Nevertheless, even the best protection is unable to prevent a component from degradation and subsequent failure. The restoration process can be applied either before the component reaches the critical level or after a failure, and its functional characteristics may be restored below or above the initial performance level.

Besides prevention from degradation, some other important properties of solid surfaces can be enhanced (Martin, 2011):

- Wear resistance
- Hardness
- Lubricity
- Corrosion and chemical resistance
- Optical properties (transmittance, reflectance, emittance)

DOI: 10.1201/9781003218654-2

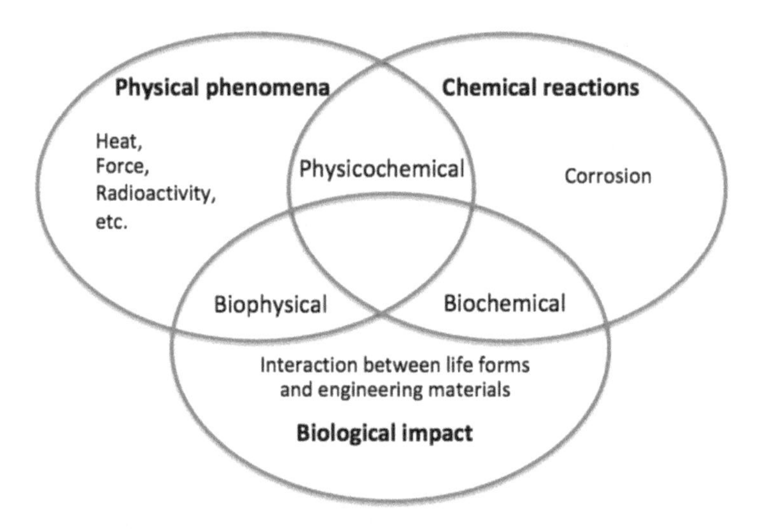

FIGURE 2.1 Main types of phenomena that cause materials degradation.

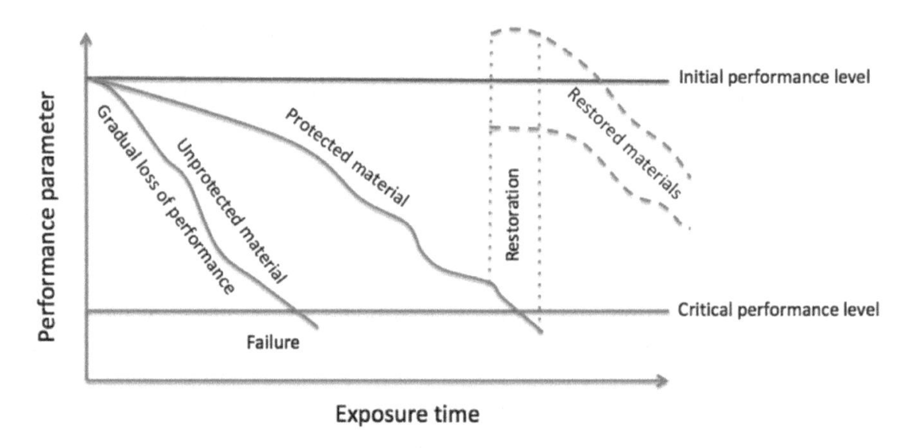

FIGURE 2.2 Graphical representation of gradual loss of performance of a component or system due to material degradation and the role of protection and restoration procedures.

- Electrical characteristics (conductivity)
- Electro-optical performance (photoconductivity, stimulated emission)
- Photocatalysis
- Surface energy parameters (hygroscopicity and hydrofobicity)
- Temperature stability

Martin (2011) lists the main objectives of surface engineering:

- Lower manufacturing costs
- Reduced life cycle costs

- Extended maintenance intervals
- Enhanced recyclability of materials
- Reduced environmental impact

From the perspective of product lifecycle prolongation and restoration of its functionality, we will address the following issues in this chapter:

- Additive manufacturing and hybrid additive-subtractive methods
- Protective and technological coatings and their properties
- Modern coating methods, such as thermal and vacuum spraying and beam coating methods
- Hard alloys and titanium nitride coatings

2.1 APPLICATION OF ADDITIVE MANUFACTURING METHODS ALONG THE PRODUCT LIFE CYCLE

A process of joining materials to make objects from 3D model data, known as additive manufacturing (AM), 3D printing (3DP), rapid prototyping (RP), and solid-free-form (SFF), is an exponentially evolving manufacturing technology (Zhang et al., 2020). Additive manufacturing (AM) with its manufacturing capabilities is becoming a key industrial process that could play a relevant role in the transition from a linear to circular economy (Sanchez et al., 2020).

2.1.1 Short Review of AM and Hybrid Methods

The widely known Additive Manufacturing is a unique manufacturing category described in the ASTM F2792 – 10: Standard Terminology for Additive Manufacturing Technologies. In general, it can be realized in three steps. First, a 3D model created in CAD software is sliced into layers. Second, the layer data in a specific format (usually standard tessellation language – STL) are given to an AM machine. The machine then adds layer upon layer of material to the substrate, thus creating 3D objects based on the STL information (Kumar and Sathiya, 2021). The AM process is shown schematically in Figure 2.3.

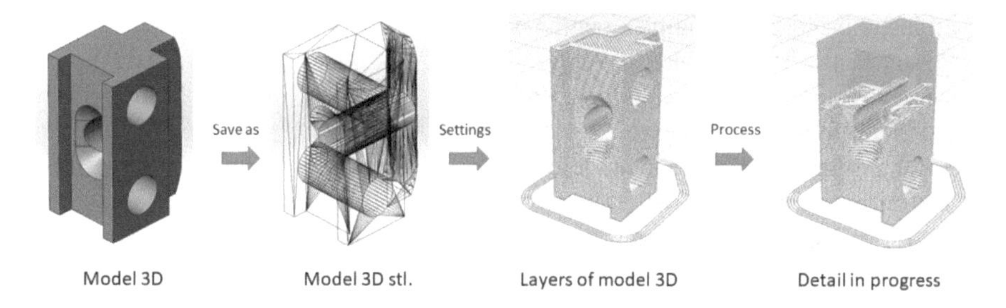

FIGURE 2.3 Process of AM techniques.

More specifically, the generic AM process can be described in eight steps (Gibson et al., 2021):

Step 1: *CAD.* A software model fully describes the external geometry. This can involve the use of almost any professional CAD solid modeling software, but the output must be a 3D solid or surface representation. Reverse engineering (e.g., laser and optical scanning) equipment can also be used to create this representation.

Step 2: *Conversion to STL.* The STL file format has become a de facto standard accepted by all AM machines, and nowadays nearly every CAD system can output such a file format. This file describes the external closed surfaces of the original CAD model and forms the basis for calculation of the slices.

Step 3: *Transfer to AM machine and STL file manipulation.* In the AM machine, there may be some general manipulation of the file to put it in the correct size, position, and orientation for building.

Step 4: *Machine setup* prior to the build process. Such settings would relate to the build parameters, such as the energy source, material constraints, layer thickness, timings, etc.

Step 5: *Building the part*, which is mainly an automated process performed without supervision. Only superficial monitoring of the machine needs to take place.

Step 6: *Removal* of the completed product. This may require interaction with the machine, which may have safety interlocks.

Step 7: *Post-processing* of the part removed from the machine, e.g., additional cleaning or removal of supporting features. The parts may require priming and painting to obtain a desirable surface texture and finish, or heat treatment to enhance their properties. Post-processing may be costly, laborious, and lengthy if finishing requirements are very demanding.

Step 8: *Application* of the parts ready for use. This may require them to be assembled together with other mechanical or electronic components to form a final product.

Additive manufacturing has important capabilities such as creating shape complexity, hierarchical complexity, material complexity, and functional complexity, so that plastic prototypes, complex engine components, houses, food, and even human organs can now be 3D printed (Wang et al., 2017). Kumar and Sathiya (2021) point out many distinctive features attainable by AM techniques, impossible for conventional manufacturing processes. First of all, AM processes are able to fabricate internal geometries of functional components and highly complex parts as a single object eliminating assembled features produced by traditional manufacturing processes, thus eliminating structural joint failures over time. In addition, direct fabrication is possible with AM processes, which involves no tooling and manufacturing sequences. Moreover, it can use functionally gradient materials and varying material compositions to produce multifunctionality features with higher mechanical characteristics. Finally, AM processes reduce material waste by recycling or reusing residual materials.

There are diverse individual processes that vary in their method of layer manufacturing depending on the material and machine technology. The range of additive manufacturing processes can be classified into the following seven categories (Singh et al., 2020):

- Vat photopolymerization
- Material jetting
- Binder jetting
- Material extrusion
- Powder bed fusion
- Sheet lamination
- Direct energy deposition

Vat photopolymerization (VPP) is a process in which a container or vat of liquid photopolymer resin is used to form an object. A layer-by-layer model is constructed with the help of ultraviolet light. This process has a high level of accuracy and gives a good finish.

Material jetting creates objects by a similar method to an inkjet printer using polymer, plastic, and a solution of sodium hydroxide. The material is jetted through a nozzle onto a previous layer surface in the form of droplets.

The *binder jetting* process uses a powder-based material as a binder, i.e., an adhesive between powder layers. The binder is usually liquid, while the build material is a powder. Metals like stainless steel as well as polymers and ceramics can be processed with binder jetting in various combinations, e.g., parts of different colors can be developed, but due to the binding characteristic this technology is not suitable for structural parts.

Material extrusion takes place when a heated material is drawn through a nozzle and then deposited layer by layer. One of the most common material extrusion processes is the fused deposition method (FDM), where a nozzle can move horizontally and a platform can move vertically after each layer of deposition. The FDM working principle is shown in Figure 2.4. The materials generally used in this process are polymers and plastics.

Sheet lamination includes two types of processes, namely ultrasonic additive manufacturing (UAM) and laminated object manufacturing (LOM). In LOM, any rolled sheet material like paper or plastic can be used effectively.

Direct energy deposition (DED) is based on a nozzle mounted on a multi-axis arm which deposits melted material on a surface where it solidifies. Generally, titanium, chrome, and cobalt are used; this process, however, is not suitable for polymers or ceramics. This technique has a great degree of control over grain structure. Zheng et al. (2021) emphasize extensive use of DED methods for repairing.

Powder bed fusion (PBF) includes all processes where focused energy (electron beam or laser beam) is used to selectively melt or sinter a powder bed layer (Neikov, 2019). Goodridge and Ziegelmeier (2017) emphasize that polymer PBF processes

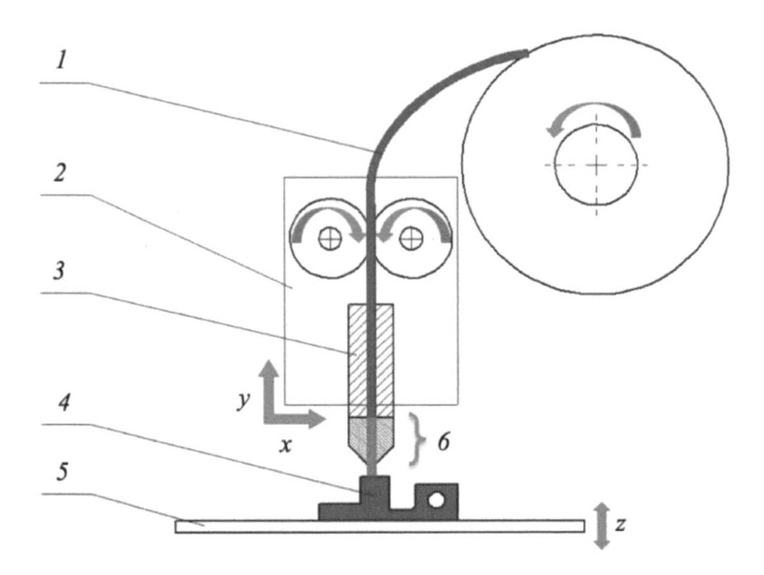

FIGURE 2.4 Fused deposition method: 1 – Filament, 2 – Extruder, 3 – Nozzle, 4 – 3D-printed part, 5 – Moving table, 6 – Heating area.

have a significant advantage over many other AM processes because they do not require support structures. During the process, overhangs and unconnected islands are supported by a surrounding unfused powder bed, which allows more complex geometries to be produced without the need for removal of support structures after the build. The parts can be stacked freely in the powder bed without supporting features, increasing the number of parts that can be produced in each build and thus increasing productivity. Some design limitations are posed by the removal of unfused powder from trapped volumes and fine channels. Another advantage of PBF techniques is the wide range of materials that can potentially be processed, which covers almost any material that can be melted and resolidified (Goodridge and Ziegelmeier, 2017).

Among all the seven above-mentioned AM techniques, PBF is one of the most widely applied (Singh et al., 2020). The powder bed fusion process is preferred because of its low-cost quality and recyclability, which is one of its best qualities. In particular, powder used in the process can be recycled to produce more parts.

In PBF, a heat source is required to fuse powder. Depending on thermal, electron, or laser heat sources available, some different types of fusion can be used in the process, including laser fusion, thermal fusion, and electron beam melting. Laser fusion can be further subdivided into selective laser sintering (SLS), selective laser melting (SLM), and direct metal laser sintering (DMLS). The SLS process is capable of fusing only plastic parts, whereas SLM and DMLS are suitable for metals (Singh et al., 2020). The principle of PBF is shown in Figure 2.5.

Selective laser sintering (SLS) is a process of combining a powder material to form a solid piece by application of heat and pressure (Singh et al., 2020). SLS is a

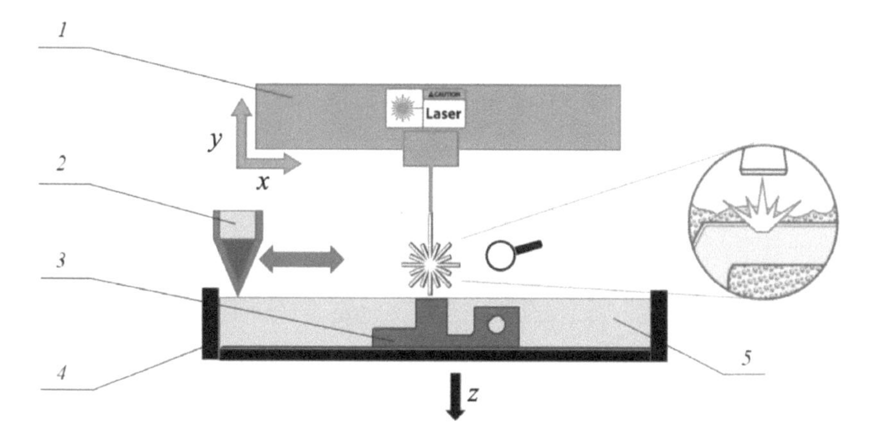

FIGURE 2.5 The basic principle of the main powder bed fusion processes: 1 – Laser unit, 2 – Powder supply, 3 – 3D-printed part, 4 – Working space, 5 – Sintered plastic powder (SLS), sintered metal powder (SLM), or melted metal powder (DMLS).

highly flexible technique which can generate high-quality complex geometry. In this method, nylon, thermoplastic, or polystyrene may be used, as well as a wide range of powder materials like polyamide (PA), glass-filled polyamide (PA-GF), and alumide. After a layer of powder is laid, a CO_2 laser performs sintering at points selected on a 2D cross section of the model (XY plane). The platform gradually descends (Z plane) in accordance with a defined layer height. The Precision of this technique can range at ±0.3% (min. ±0.3 mm), the minimum layer thickness is 0.08 mm, and maximum model size 700 × 380 × 580 mm. The process is fast and accurate, a superior-quality surface is obtained with a minimum material wastage. There is no need for mold or other tools or fixtures (Singh et al., 2020).

Electron beam melting is one of the latest technologies, where the energy source for the melting process is an electron beam emitted from a tungsten filament and controlled by a coil. Fabrication of titanium parts is possible and a layer thickness of 0.1% can give better results in a shorter time, reducing the cost by up to 35% (Singh et al., 2020). The idea of the electron beam melting AM process is shown in Figure 2.6.

The selective laser melting (SLM) principle is similar to SLS fabrication: layer by layer with the help of high-energy laser beams on a powder bed, but the powder is melted rather than sintering. The minimum thickness of the powder material layer is 0.020 mm. SLM is faster than SLS, but it requires inert gases for the laser. It is currently a very popular method for fabrication of metal parts (Singh et al., 2020).

Direct metal laser sintering (DMLS) is developed with the use of metal (not plastic) powders, e.g., titanium and its alloys. Similar to SLS, it is a layer-by-layer process with a computer-controlled laser technique. An object is fabricated on a movable platform by applying incremental layers of a pattern material. Each layer has an equal thickness of approximately 0.1 mm. The DMLS method makes it possible to control porosity of each layer, which cannot be eliminated completely, though (Singh et al., 2020).

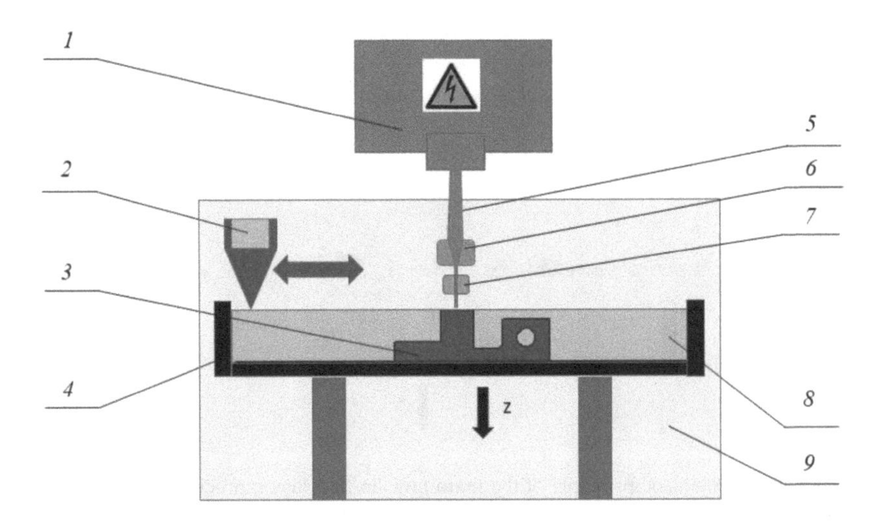

FIGURE 2.6 Electron beam melting: 1 – Electron gun unit, 2 – Powder supply, 3 – 3D-printed part, 4 – Working space, 5 – Electron beam, 6 – Focusing coil, 7 – Directing coil, 8 – Powder bed, 9 – Vacuum chamber.

It should be noted that the powder bed fusion technique has also been applied to part repairing. In comparison with DED, the capability of PBF for remanufacturing is restricted because of smaller build envelopes and limited accessibility of new material deposition (Zheng et al., 2021). For instance, the material can only be deposited on a flat surface but not inside a concave structure.

According to Kumar and Sathiya (2021), the AM processes can be categorized based on the state of the starting material, namely liquid, filament/paste, powder, and solid sheet:

In the *liquid state*, such materials as photopolymers, especially acrylate-based ones, hydrogel, or inert urethane wax can be processed with SLA-based methods, nonlithography 3D printing, or multi-jet fluid dispensing.

Thermoplastics, ceramics, and polymer integrated with metal particles can be processed in the form of a *filament or paste* using FDM machinery.

As *powders*, metals, ceramics, and polymers can be starting materials for laser-based AM processes, and especially magnetic alloys can be formed with laser engineered net shaping (LENS) methods.

Finally, *sheets* of metals, ceramics, or thermoplastics can be used in LOM processes to form 3D objects.

An important aspect of metal additive manufacturing (MAM) is hybridization with material removal processes, the so-called 'subtractive processes.' Pragana et al. (2021) grouped them into two different categories:

1. Utilization of material removal processes at *post-processing level*, usually performed in order to obtain higher geometry precision, dimensional tolerances, and surface quality. This category can be considered the simplest,

but nevertheless very important in lowering overall material and energy consumption. This category of methods is recognized in the case of processing as expensive and difficult-to-work materials such as titanium or nickel-based super alloys as a successful way to reduce the buy-to-fly, which is the ratio of the initial workpiece mass to the mass of a finished part. Hybridization of metal additive manufacturing with subtractive processes at post-processing level is very effective at lowering overall manufacturing cost and material waste.
2. Additive/subtractive processes, i.e., integration of *material removal processes with material addition during a manufacturing sequence.* This way, parts can be obtained that would be almost impossible to produce either by additive manufacturing or by material removal operations. An example of such a hybrid process is shown in Figure 2.7.

Hybridization of MAM with forming processes is an important direction. Here, medium or large batches of semi-finished parts are produced by forming, while functional elements are subsequently added by additive manufacturing (Merklein et al., 2016). This sort of hybridization of metal additive manufacturing (metal HAM) is recognized as an effective approach to extending conventional forming process routes to fabrication of tailor-made, customer-oriented products, to improve properties of the deposited metals both during and at the end of a process route, and in particular to ensure defect-free flow and die filling with minor metal losses during small-batch, single-stage forming operations (Pragana et al., 2021).

Taking into consideration the above-mentioned characteristics, Pragana et al. (2021) decided to group the combination of MAM with forming processes into four categories:

1. Integration with processes to improve properties of the deposited metals
2. Integration with bulk forming processes

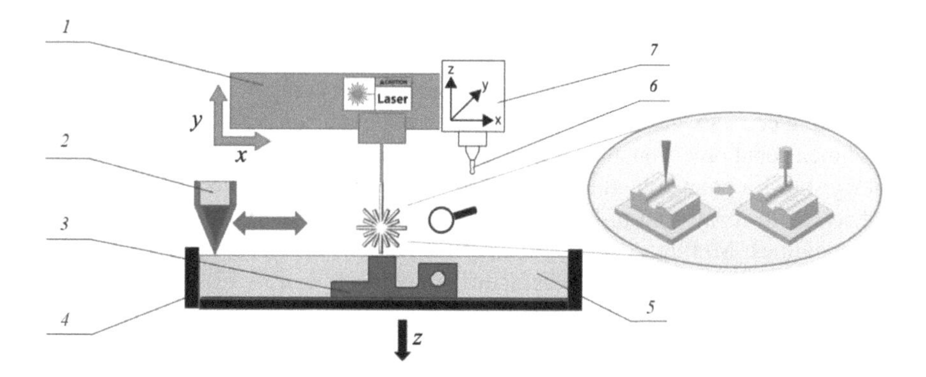

FIGURE 2.7 Example of hybrid additive-subtractive processes (milling combined with PBF): 1 – Laser unit, 2 – Powder supply, 3 – Workpiece, 4 – Working space, 5 – Powder bed, 6 – End mill, 7 – Spindle.

3. Integration with sheet forming processes
4. Integration with joining by forming processes

The roots of combining metal HAM with processes *to improve properties of deposited metals* can be found in the processes of mechanical surface treatment, such as application of pressure along weld beads produced by friction stir welding as a means of controlling residual stresses and distortions. This process, known as surface rolling, subjects weld bead surfaces to plastic deformation in order to improve surface finish and induce compressive stresses that will counteract residual stresses originating from heating-cooling cycles of welding. Another hybrid method in this group is utilization of shot peening on successive layers of wire-arc additive manufacture (WAAM)-based processes for relieving residual stresses and minimizing distortions and to improve fatigue life due to the effect of compressive stresses on delaying initiation of fatigue cracks. The combination of MAM with hot forging through utilization of a customized WAAM torch may become an alternative solution for decreasing residual porosity, refining microstructure, and improving mechanical properties of a deposited material.

Integration with bulk forming processes in the case of an aluminum alloy AA5083 deposited by the WAAM method achieved an excellent ductility of deposited material. The combination of additive manufacturing with coining has also been recently proposed as a novel process route to fabricate high value-added collector coins.

The *integration of metal additive manufacturing with sheet forming processes* comprises combinations of MAM with bending, deep drawing, spinning, and incremental sheet forming.

Finally, *integration with joining by forming processes* found its application in the area of assembly of structural components. It needs to replace conventional welding, fastening, and adhesive bonding as it eliminates metallurgical problems caused by heating–cooling cycles and by incidence of hard and brittle intermetallic compounds with the aptitude to join dissimilar materials. Hybridization of additive manufacturing with joining by forming processes provides flexibility, environmental compliance, and adequacy for producing small, medium, and large batches of joints (Pragana et al., 2021).

Zheng et al. (2020) stress that hybrid manufacturing combines advantages of AM in building complex geometries and subtractive manufacturing's benefits in obtaining dimensional precision and surface quality. The hybrid additive-subtractive technology shows a great potential for supporting repair and remanufacturing processes and serves to repair end-of-life parts or remanufacture them to new features and functionalities. Mixing AM and subtractive manufacturing provides a more flexible, productive, and capable manufacturing approach, and increases the ability to remanufacture to a higher standard (Zheng et al., 2020).

2.1.2 Materials Processed by AM

A broad range of materials has been developed for AM techniques, including polymers, metals, ceramics, glasses, biomaterials, and multi-material systems. Further improvement to functionality of material structures can be achieved through hybrid or multi-process 3D printing (Liu et al., 2021). However, additive manufacturing of pure metals is difficult since they exhibit relatively poor mechanical properties (Cooke et al., 2020).

According to Sanchez et al. (2020), *plastics and polymer materials* are a key area in the field of AM, by far the most used material type. They include thermoplastics, thermosets, elastomers, hydrogels, functional polymers, polymer blends, composites, and biological systems. Most amorphous, thermoplastic materials are processed by material extrusion methods, with fused filament fabrication (FFF) and fused deposition modeling (FDM) being the most popular techniques. In these techniques, polylactic acid (PLA) and acrylonitrile butadiene styrene (ABS) are among the most widely used materials, but general polymers that can be melted at an adequate temperature without degradation can be processed with material extrusion systems. Technical requirements of materials include interfacial adhesion and undisturbed polymer entanglement to allow for manufacture of nonporous objects with mechanical properties similar to products made by conventional techniques. In particular, rheology, thermal, and mechanical properties need to be characterized to validate a candidate material for this application, such as geometric characteristics, tensile, fatigue, flexion, etc. (Sanchez et al., 2020).

Cooke et al. (2020) point out that the metal AM industry has seen a surge in 3D printer technology with highly advanced devices and methods, but the number of metal alloys that can be reliably printed is minimal. A vast majority of metal alloys used in conventional processes are unable to produce adequate parts in AM. Among *metallic materials* that can be used in AM techniques are steels, aluminum (Al) and magnesium (Mg) alloys, titanium (Ti) and its alloys, and Ni/Co-based alloys. Kumar and Sathiya (2021) provided their short review as follows.

Only a tiny fraction of available Fe alloys can be processed with AM methods. Different *steel materials* such as stainless steel (SS), austenitic stainless steel, maraging steel, and tool steel are generally used in both general and tooling applications where high hardness and strength are required. Some steels, e.g., hardenable SS and austenitic steels, are sensitive to parameters of AM techniques. For machine elements, fasteners, and tools, 34CrNiMo6 steel with high strength, toughness, and hardness is used. However, when this material is processed with a laser, significant changes in its microstructure and mechanical properties take place. After processing 300 M steel by laser solid forming (LSF), as-deposited martensite and coarse bainite turn out to be uniform during heat treatment. Uniform transition of tempered martensite and bainite retains a small amount of austenite. These microstructural changes may cause fatigue crack to grow.

Al and its alloys are very attractive for aerospace engineering due to their light weight and appreciable mechanical properties. They are available at low cost and, having high thermal conductivity, they enable quick fabrication with reduced thermal stresses compared to other metallic materials. However, high-performance Al alloys have poor weldability and are highly reflective in the range of laser wavelengths, which pose some limitations to their applications. Among the commonly used Al alloys are AlSi10 and AlSi12. The Al alloy with added scandium fabricated with the laser-based AM process exhibited the highest tensile strength of 530 MPa along the build orientation and a remarkably elevated hardness after aging treatment, revealing a potential to become a highly demanded material for aerospace applications.

Due to their low weight and promising strength characteristics, *Mg and its alloys* are widely used by aeronautical industries tending to replace heavier Al alloys. Mg alloys are also used as an alternative replacement for plastics, since they have more stiffness than and density comparable with plastics. Additionally, when Mg alloys are doped with rare earth elements, they show an enhanced corrosion resistance as well. It should be noted that processing of Mg alloys brings the risk of firing when a material is a powder or at higher temperatures, which limits the use of Mg alloys in aerospace industries. Nevertheless, different AM techniques can be adapted for fabricating magnesium components and medical implants with a reduced ignition risk, e.g., PBF methods, WAAM, FDM, friction stir AM, and binder jetting techniques.

Ti alloys are widely used in aerospace and biomedical applications due to their predominant properties, such as higher specific strength and outstanding corrosion resistance. These superior properties enable Ti alloys to be used for high-temperature and strength applications such as blades and cases of steam turbine and turbo engines. While manufacturing of Ti alloy components with conventional techniques is challenging due to their low thermal conductivity and affinity with cutting tool materials, laser-based AM techniques are effective methods of Ti-based alloys processing. In particular, electron beam AM techniques can produce Ti alloy parts of higher densities than obtainable by SLM. There are a few Ti-based alloys recently of interest to researchers, such as Ti6Al4V, Ti8Al1Er, TC11, TC21, and Ti5553, commonly used for AM fabrication of components. In particular, Ti6Al4V alloy, known also as Ti64 alloy, is a widely accepted $\alpha+\beta$ Ti alloy that accounts for nearly half of the Ti market share around the world. Another alloy processed with laser-based AM is Ti-6.5Al-3.5Mo-1.5Zr-0.3Si, sometimes referred to as TC11, which is widely used in compressor disks and other aerospace applications due to its capacity to withstand high strains during loading.

Ni-based superalloys such as Inconel 625 and 718 are primarily used to fabricate critical components, such as gas turbine engines and compressors, that regularly undergo high temperatures and stresses. They can be fabricated by the EBM or laser-based AM processes, as well as binder jetting in the case of the Inconel 718 alloy. On the other hand, *Co-based alloys*, e.g., CoCr, are used for dental and other biomedical applications. The AM-prepared Ni–Co-based alloys reportedly produce synchronous improvements to both microstructure and mechanical properties as Co inclusion rises. AM of *Ni-based shape memory alloys* (SMAs) is another exciting, rapidly developing research field. These alloys can be deformed in cold conditions, but when they are subjected to heat, the element regains its pre-deformed shape due to a phase transformation induced by heat. NiTi and Cu–Al–Ni are considered to be the most common SMAs used in aerospace applications, biomedical applications, and conventional actuators for pneumatic, hydraulic, and motor-based systems (Kumar and Sathiya, 2021).

As emphasized by Cooke et al. (2020), metal AM systems require a metal feedstock to be melted and deposited on a substrate to then construct layers which eventually form a desired geometry. Some systems use a solid wire feed, most commercial methods employ metal powder. The two most common commercialized methods for MAM include powder bed fusion (PBF) and direct energy deposition (DED) with laser powder bed fusion (LPBF) printers being the most widely available systems on the market. LPBFs attain a superior resolution with a typical layer thickness between 10 and 50 µm, while powder-fed DED printers have around a 250 µm layer thickness. However, an important factor in both precision and quality of the build begins with preparation of the starting alloy powder. Consistent powder geometry, composition, and flowability are required to ensure repeatability of metal printed structures. A specific size of particles must ensure a good packing behavior and limit porosity. The average particle size distribution is 10–45 µm in LPBF and 20–200 µm in DED. Powder flowability improves with increased sphericity, which is reflected in quality of a finished product; but if the powder is too flowable, then spreading defects can be produced during deposition. If metal powder properties such as size, sphericity, surface texture, and chemical composition vary within a single batch, a printed part may be produced with more prominent defects, such as high porosity and poor surface finish.Cooke et al. (2020) present several techniques for atomizing metal for PBF and DED applications, including gas atomization, high-pressure water atomization, plasma rotating electrode, plasma atomization, electrolytic processing, and mechanical crushing. They point out gas atomization and sometimes high-pressure water atomization are the most common and preferred methods in the industry. Both the processes pour molten metal through a nozzle. When it comes in contact with a high-pressure fluid stream, the stream separates and quenches the metal into small particles which can then be used as a starting powder for MAM (Cooke et al., 2020).

Before describing application of *structural materials* to additive manufacturing, Liu et al. (2021) introduce the concept of four-dimensional (4D) printing, which has emerged with involvement of versatile shape-morphing systems. Typical examples of shape-morphing assemblies are origami and kirigami, the former being the art of folding thin sheets into 3D objects with rich geometric algorithms, and the latter a variation of origami in which a material is cut when the structure is folded. In 4D printing, certain environmental stimuli, such as exposure to heat, magnetic fields, liquids, electricity, light, gas, prestress, or their combinations are posed on a 3D-printed material. As a result, in response to these stimuli, the material autonomously and programmably changes its configuration or function. With 4D printing, a multi-material strand can be folded into a desired shape and in the last decade various smart materials have been developed for 3D printing and self-shaping assembly. Multiple prospects for structural materials AM have been raised, including multi-material AM (MMa-AM), multi-modulus AM (MMo-AM), multi-scale AM (MSc-AM), multi-system AM (MSy-AM), multi-dimensional AM (MD-AM), and multi-function AM (MF-AM) (Liu et al., 2021).

Additively manufactured materials, apart from a microstructure typical for each respective AM method, may possess some *process-induced defects*. In a part produced by PBF, defects are often similar to those after welding (Singh et al., 2020). Defects like cracking, high surface roughness, unintended anisotropic mechanical and physical properties, or anisotropic shrinkage depend on controllable process parameters having a complex influence over thermal behavior of the melt pool. One of the most common factors for these defects is wrong heat input rates during high-temperature solidification. Surface defects usually occur because of high cooling rates and repeated thermal loading. Speidel et al. (2021) named other defects induced by PBF processes that are widely used, such as poor surface quality generated by layer-by-layer supply, unfused powder, balling, and subsurface porosity. The authors emphasize that a combination of discretely sized powder feed-stocks influencing roughness and the very layer-based manufacturing approaches leading to staircase effects mean that *most AM parts require some form of post-processing and surface finishing* prior to application.

Among the post-processing techniques, metallurgical methods with hot isostatic pressing should be mentioned, as well as principal heat treatments and stress relief methods (Singh et al., 2020). In the heat treatment process, it is very important to obtain a proper microstructure that affects mechanical properties of metal. The thermal treatment is required also in order to reduce porosity and other defects. In the case of EBM parts, due to the higher processing temperature, there is little need for stress relief, but some material may still benefit from HIP in the heat treatment process to get a more optimal microstructure (Singh et al., 2020).

Solution, Quench, Age, and Temper Heat Treatments. For precipitate hardened materials, a solution heat treatment is required. It should be long enough to dissolve the precipitate but short enough to limit grain growth. These processes of solution, quench, age, and temper heat treatment tend to be alloy specific, but they give different starting conditions of AM materials compared to cast and wrought materials of the same composition (Singh et al., 2020).

Hot Isostatic Press (HIP). Most metals are heat-treated by this method to close down pores and cracks developed in the material. This process involves pressure and high temperatures working simultaneously to reduce porosity that may lead to a poorer tensile and fatigue performance. Most post-processed AM parts can undergo HIP treatment. Although it is not necessary to treat all additively manufactured parts this way, when high performance or safety is required, HIP is included in the standard process for a wide variety of materials (Singh et al., 2020).

Stress Relief. It is the process of material recovery. At an elevated temperature, the atomic diffusion rate increases and atoms from the region of higher stress start to move toward the region of lower stress, which results in relief of internal strain energy. PBF produces a high thermal slop layer by layer, which generates a high degree of localized residual stresses throughout the part, thus a stress relief is necessary to equalize the stress gradients (Singh et al., 2020).

2.1.3 AM Methods Related to the Product Life Cycle

Sanchez et al. (2020), trying to identify sustainability benefits of AM, distinguish four main stages of a product lifecycle:

1. Design
2. Production

3. Consumer/Prosumer
4. End of Life

The authors state that at the *design stage*, a clear advantage of AM in environmental aspects is the opportunity to produce more complex and optimized components with reduced number of joining and assembling operations. Its higher flexibility than in traditional manufacturing improves the product development cycle, reducing the overall time and cost and improving human interaction. However, the geometric freedom of products is limited by operational requirements and process constraints. In some cases, as indicated by Garashchenko and Rucki (2020), a part decomposition procedure must be performed in order to ensure efficient use of the AM machine workspace and to provide savings of material and energy.

In the *production phase*, three main aspects should be considered (Sanchez et al., 2020): (1) resource consumption, (2) waste management, and (3) pollution control. In respect of the resource consumption, the energy consumed by AM equipment and auxiliary subsystems and material consumption are lower than that of traditional processes. Considering the waste management, the layer-by-layer manufacturing principle certainly improves material yields, i.e., ratio of final product weight to input material weight. Additive manufacturing seems to represent an opportunity to implement the roadmap to zero waste manufacturing through development of direct digital manufacturing. However, examples of waste include unused and unusable powdery materials, waste generated by unexpected defects, and supporting structures necessary in some 3D printing processes. Finally, as far as the pollution control is concerned, AM uses fewer auxiliary harmful chemicals than conventional manufacturing, where forging lubricants, cutting fluids, or casting release compounds are usually necessary.

In the *consumer phase*, adoption and diffusion of additive manufacturing by different communities have resulted in a growing interest in personal fabrication, do-it-yourself (DIY), and peer-to-peer practices in open spaces (Sanchez et al., 2020). It can be stated that capacities of digital manufacturing are undergoing a democratization process and widespread use of the technology, which allows private and industrial users to design and produce their own goods in the framework of the 'prosumer' concept (the merged words 'producer' and 'consumer' underline the changing perspective of a more home-focused general economy). In the *prosumer context*, four drivers are to promote environmental sustainability in personal fabrication: (1) product longevity, (2) co-design, (3) local production, and (4) technology affordance. Nevertheless, lack of knowledge about AM has some impact on the social and environmental aspects, since AM practitioners might not be aware of the environmental dimensions when choosing an AM technology (Sanchez et al., 2020). It can be added after Pfähler et al. (2019) that within the post-production stage, during product use and application, associated with service and reconditioning, AM can be used for manufacturing of spare parts and repair of wearing parts well before they reach EoL stage.

The *end-of-life (EoL) stage* attempts to close the product lifecycle loop and in AM it can be achieved at different levels. A key feature is repair and maintenance ability that AM could contribute to in order to extend the product life span. This cost-effective approach has been usefully exploited for metal parts offering a great potential for repair of damaged components. Likewise, reverse engineering is an approach to foster repairing and refurbishing. Finally, concerning the recycling process, several initiatives have been reported to create low-cost extruders to produce plastic filament for FFF devices. Recycled filaments and some organizations recycling waste plastic toward products with a higher added value can be mentioned as important recent trends (Sanchez et al., 2020).

Le et al. (2017) performed a review of AM techniques in repairing and remanufacturing. They provide examples of laser cladding used for building of new Ti-6Al-4V entities on old Ti-6Al-4V substrates and of directed energy deposition (DED) technologies especially suitable for repairing and remanufacturing as well as adding new functionalities to existing parts. Due to the flexible material deposition configuration of a five-axis CNC machine, these techniques have been successfully applied to remanufacturing of worn-out or damaged components, particularly for high-value components, such as turbine blades, molds, and dies, restoring them to original specifications and qualities.

The authors point out that powder bed fusion (PBF) techniques, such as electron beam melting and selective laser melting, have some limitations in repairing and remanufacturing abilities, compared with DED, due to their limited build envelope and deposition of materials on horizontal flat surfaces only. However, there are numerous components which can be repaired or remanufactured by these processes, e.g., gas turbine burner tips. EBM technology enabled building a copper entity on top of an existing Ti-6Al-4V part with a good metallurgical bonding at the interface between the two materials. It was found that EBM not only has the potential to produce multi-material parts, e.g., joining Inconel 718 with 316L Stainless Steel, but also to be used for remanufacturing applications and building new features on existing parts. The existence of strong bonding between EBM-built entities and the existing part was demonstrated by microstructure observation and tensile testing. In a similar way, multi-material parts can be built by SLM ensuring a good bonding at the interface between two materials, while new features can be built on existing components. This way it is possible to obtain new parts by adding new entities to the existing components. The parts obtained have a good "material health", which means that their mechanical characteristics are compatible with industrial applications (Le et al., 2017).

Pfähler et al. (2019) present results of a survey of AM's application possibilities considering product life cycle. They mention rapid prototyping, rapid tooling, rapid manufacturing, and efficient product as well-established applications of AM within the manufacturing industry. In particular, *rapid prototyping* is used for rapid production of prototypes, where AM is most likely to be used for fast production of small batches on demand reducing the time to market. *Rapid manufacturing* describes usage of AM for large series and small series production which is achieved through the increasing diversity of materials available in AM.

Rapid tooling describes manufacture of tools and appliances for production. When a small quantity of components is required and production of a required tool is not economical, AM can be used for rapid production of this tool (Pfähler et al., 2019). On the other hand, AM technologies enable producing tools of complex geometry, for example, a drill base body with coolant ducts shaped spirally along flutes without weakening the drill core (Tyczynski et al., 2019). This is related to the *efficient product* concept, which stands for implementation of additional product utility which is made possible by AM through, for example, freedom of design (Pfähler et al., 2019).

Pfähler et al. (2019) also note *production of spare parts* on demand, which is becoming an increasingly relevant application area of AM. Spare parts can be manufactured directly when and where required, avoiding unfavorable storage costs and eliminating the need for transportation. Moreover, AM can be used to *repair wearing parts* instead of replacing an entire component with a new one. An area that needs to be repaired can be fixed using an additive technology, e.g., laser deposition welding.

Concluding their survey, Pfähler et al. (2019) observe that a majority of participating companies are active in areas like rapid prototyping, rapid tooling, rapid manufacturing, and efficient product. Considerably fewer companies are currently active in the remaining fields of application such as repair of wear parts and spare parts on demand. However, the authors believe that the two latter categories show notable potential for future usage.

Le et al. (2017) describe two potential scenarios for manufacturing parts from existing components using AM technologies. *The first scenario* included hybridization of additive and subtractive processes, as well as the inspection process. In this scenario, an existing used-up part is utilized again directly as a workpiece to produce a new part or final part. The metal powder required to produce the final part is made of raw material by the gas atomization process, while chips generated in the subtractive phases of manufacturing are recycled into a raw material. Sintered and/or non-melted powders in AM processes are also recycled for reuse in the next production cycle. *As part of the second scenario*, an existing part is first recycled into raw materials which will be used for producing workpieces by conventional methods such as casting or forging, and the final part is made from the workpiece by a sequence of machining and inspection operations. In this scenario, the chips are also recycled into a raw material for a new production cycle. In both scenarios, final inspection is applied to verify the required quality. If the dimensions of manufactured part are out of tolerances, a decision will be made to continue processing the part according to the first or second scenario.

In their study, Le et al. (2017) propose a direct manufacturing approach according to the first scenario. Utilizing benefits of combined additive and subtractive technologies, this approach can give a new life to existing metallic or non-metallic parts by transforming them into final parts intended for another product. The authors focus on the metal-based AM techniques, in particular, DED and PBF, and existing parts as components extracted from an EoL product. The existing part has been identified in terms of material quality, size, and shape, which are compatible to produce a final part of different features. This enables not only reusing material of an existing part effectively, but also avoiding the material recycling phase and several subtractive and additive operations to achieve the geometry and quality of the final part intended for another product. This way, EoL parts were given a new life in their life cycle. The proposed approach seems to have a potential to reduce energy and resource consumption as well as waste during the manufacturing process. Unlike the traditional remanufacturing of components, the proposed approach transformed EoL parts into ready-to-use parts with new functionalities, totally different from those of their previous application, intended for another product and new usages (Le et al., 2017).

In the framework of AM related to the product life cycle, Sanchez et al. (2020) describe the concept of distributed recycling via additive manufacturing (DRAM). It is related to use of recycled materials by means of the mechanical recycling process in the additive manufacturing process chain that consists of six stages, namely recovery, preparation, compounding, feedstock, printing, and quality inspection. The results of their survey suggest that activity at the recovery and preparation stages is limited, while at the remaining stages important achievements are seen in validation of technical feasibility, environmental impact, and economic viability.

AM technologies gain additional significance in the context of plastic waste pollution that poses a major threat because of non-degradability affecting ecological environments. Sanchez et al. (2020) underline that recycling rates of plastic packaging remain small on a global scale, approx. 14%. The European Commission identified dealing with plastic materials as a priority area, with the aim of making all plastic packaging recyclable by 2030. Regarding industrial ecology of polymers, four main approaches to recycling plastic solid waste can be identified as primary, secondary, tertiary, and quaternary recycling. The primary and secondary recycling is performed as a mechanical recycling process, but for complex and contaminated waste, chemical recycling is a preferable option. In order to use high-grade recycling in the circular economy framework, four conditions need to be met (Sanchez et al., 2020):

1. An adequate collection system and logistics
2. Guaranteed volumes of supply
3. Market demand for recycled materials
4. Quality guarantee of recycled materials

The authors point out three levels of opportunities for DRAM driving the circular economy:

1. **Micro level**
    - Development of low-cost, free and open source, digitally manufactured (ideally from recycled waste) tools to enable distributed DRAM including tools for grading and typing recycled waste plastic, tools for separating and shredding materials, and tools to either produce waste plastic filament or directly 3D print waste plastic.
    - Development of novel waste-based composites that involve material characterization, as well as life cycle economic and environmental assessments of DRAM.
    - Development of applications of the above-mentioned new DRAM waste material composites and markets for these applications.

2. **Meso level**
   - Finalization of complete closed-loop DRAM case studies based on material and location to assess technical, ecological, and economic feasibility.
   - Development of business models that can fabricate and sell these DRAM tools, components, and services around calibrating, using, and maintaining them.
   - Development of paths to enable existing recycling organizations and businesses to convert their activities to these new DRAM-focused business models.
3. **Macro level**
   - Development of policies to provide incentives for open-source development of the above-mentioned paths for DRAM.
   - Development of educational materials and school programs to implement DRAM in public and private schools, community centers, etc.
   - Development of means to disrupting fossil-fuel-based plastic markets by offsetting materials with DRAM-based products (Sanchez et al., 2020).

2.2 PROTECTIVE AND TECHNOLOGICAL COATINGS FOR HOT WORKING PROCESSES AND HEAT TREATMENT

During materials processing, when steel components are heated in the presence of air or products of combustion, oxidation and decarburization phenomena take place. These can lead to quality impairment through dimensional changes, worsened surface finish, quench cracking, etc. Measures against scaling and decarburization are undertaken in many hot working technologies and heat treatment. It was reported that decarburized depth of heated constructional and alloyed steels can be reduced to 6–10% by means of protective coating (Alexenko et al., 2016). In fact, it can be assumed that protective technological coatings can be useful at any stage of the manufacturing process from the ingot to ready-to-use components and units.

2.2.1 General Requirements

Coatings used for protection of hot processed workpieces differ from those designed to protect finished parts during their operation in terms of both chemical composition and physical characteristics. Any coating is expected to protect the outer surface of a component, but in the case of temporary technological coatings it is especially important to prevent scaling in the heating process. Thus, a technological coating should form a continuous and thermally stable film, isolating a heated metal from the environment. When protecting high-performance steels and alloys, technological coatings must shield the surface from decarburization, burnout of alloying elements, intergranular gas corrosion, and contamination with interstitial elements, such as oxygen, nitrogen, or hydrogen. The temperature range of protection is very wide, from 500 to 2000°C, and duration of hot processing can span between several dozen seconds in the case of induction heating to dozens of hours in flame furnaces. Depending on chemical composition, technological coatings should be able to melt quickly and form protective films, as well as maintain their viscosity in long-term heating.

In hot forming processes, a coating should display a required viscosity and form a thermoresistant technological lubricant. It should not interrupt plastic deformation of the hot metal, sustain high pressure and temperature, and make it easier to remove the workpiece out of the die. Higher levels of protection are achieved at better wettability and adhesion to the component surface. Coatings are often expected to be self-removable and crush out or peel off after a technological operation.

Moreover, a technological coating should be non-toxic, incombustible, and easily available, and should ensure its low consumption per unit area of a protected metal surface.

2.2.2 Starting Materials for Coatings

For protective technological coatings, a variety of materials can be used of different natures, chemical compositions, and properties. Considering the wide range of the hot working processes in various temperature and time conditions, as well as effects of furnace design on heating rates, the number of different coating materials should be very large. These may be categorized according to their nature, since ceramics and glasses are radically different materials than metals, but are close cousins to each other, though ceramics are crystalline, while glasses are amorphous (Messler, 2006). Hence, glasses progressively soften upon heating and never melt, while ceramics almost always exhibit high melting temperatures and/or thermal stability. Thus, these three categories may be applied to classification of coatings:

1. Glass-like materials and glasses
2. Ceramics and pure oxides (Al_2O_3, ZrO_2, MgO, Cr_2O_3)
3. Metals and intermetallic compounds, carbides, nitrides, etc.

For *glass processing*, the starting material is a glass batch, which is formulated from a variety of materials, some mined from the earth and used with only a few preparatory steps, and some more refined, such as metal oxide powders (Francis, 2016). Materials in a glass batch perform different functions, either supply a metal oxide for glass composition or are additives that help produce a uniform melt. Glass powder or frit is considered a glass starting material and is prepared by quenching a uniform glass melt. A glass batch is heated in a furnace and converted into a homogeneous melt. Quenching a glass melt from a high temperature into water or between metal rollers results in thermal shock, so that glass is fractured into fragments, which further can be broken down using ball milling to create a frit of a desired particle size. When heated, a powdery glass batch is converted to a homogeneous glass melt (Francis, 2016).

Vitreous coatings are made of enamel frits that consist of alkaline alumina-borosilicate to which other inorganic substances may be added to provide desirable physical properties, while silica is the fundamental oxide of glass (Hossain et al., 2014). Sitall coatings are used to ensure specific characteristics, such as heat-resistant and high-temperature corrosion protection (Lazareva, 2020). Ground-coat frits contain more boron oxide than the enamels that are used for the cover coat, while alumina can be added to control viscosity, surface smoothness, and mechanical hardness of the glass (Schäfer, 2010).

Glaze surface quality, its microstructure, and physicochemical properties depend on its composition and production method. Effects of various corrective additives and most frequently used oxides on glaze properties are described by Gerasimov and Spirina (2004):

- Quartz (SiO_2) is introduced in the form of quartz sand, pegmatite, and as part of clay and kaolin. It raises melting point and viscosity of a melted glaze, improves acid resistance, mechanical strength, and luster, and decreases the thermal coefficient of linear expansion (TCLE).
- Alumina (Al_2O_3) introduced via clay, kaolin, feldspar, and calcined alumina raises refractoriness and viscosity of a melt, improves chemical resistance, reduces the TCLE, and impedes good spreading of fritted glaze.
- Boron oxide (B_2O_3) is introduced in the form of borax, boric acid, and calcium borate. Its presence significantly decreases the melting point and the TCLE, increases leachability, improves luster, and facilitates good spreading.
- Lead oxide (PbO) is an intense flux added in the form of minimum white lead and lead monoxides. It facilitates expansion of melting interval, imparts a specially attractive luster, facilitates good spreading of glaze, and decreases its hardness and TCLE.
- Strontium oxide (SrO) is a flux, too. It is used in low-melting glazes instead of PbO, has a positive effect on the firing interval, hardness, and chemical resistance, and improves luster. It is introduced with strontium carbonate.
- Sodium oxide (Na_2O) is an intense flux, which raises the TCLE, decreases fusibility and viscosity interval, lowers hardness, and improves the luster of glaze coating. It is introduced into batch in the form of soda, borax, and sodium feldspar.
- Potassium oxide (K_2O) is an intense flux as well. It is introduced via orthoclase and potash and increases the fusibility interval, improves luster, and increases the TCLE. Potassium oxide is preferred to sodium oxide.
- Lithium oxide (Li_2O) is the most intense flux among the alkali oxides. It is introduced into a batch via spodumene and lithium carbonate and decreases viscosity, improves luster, and increases the TCLE.
- Calcium oxide (CaO) is the main flux for medium- and high-melting glazes, since its effect is weakly manifested in glazes with a melting point below 1150°C. Calcium oxide cuts viscosity of a glaze melt with a high content of SiO_2 and increases hardness and strength. It is introduced in the form of chalk, marble, and opoka.

- Magnesium oxide (MgO) is a stronger flux than CaO. It increases hardness, luster, strength, and elasticity of a glaze coating. It is introduced in the form of opoka, talc, magnesite, and dolomite.
- Barium oxide (BaO) is a flux introduced in the form of barium carbonate. It decreases the TCLE and, when introduced in a small amount, improves luster and mechanical properties of a glaze.
- Zinc oxide (ZnO) is a good flux. It decreases the TCLE and facilitates crystallization and opacification of a glaze. It is added in the form of zinc white.
- Zirconium oxide (ZrO_2) is an opacifier that increases hardness and chemical resistance. It is introduced in its pure form or as zircon, which is itself an opacifier.
- Tin oxide (SnO_2) is a better opacifier, increasing chemical resistance, especially alkali resistance.

As raw materials for technological temporary coatings, volcano glass and wastes from glass production can be used. Addition of broken, used-up glass appears to be very effective since it avoids melting of frits. Application of frits mixed with broken EoL glass, volcanic ashes, and metallurgical slag seems to gain in popularity.

It should be noted that heat-resistant vitreous coatings of finished products may be single-component or complex with a higher softening temperature than normal enamels. Crystalline coatings may be either entirely crystalline, composite, or glassy-crystalline. Composite high-temperature coatings are a combination of a glass matrix and refractory filler. They exhibit good operating and physicochemical properties as a result of combining the composition of multicomponent silicate enamels and refractory or chemically stable fillers, and they are used to protect low-carbon, unalloyed, and alloy steels, nickel, chromium–nickel, and titanium alloys from high-temperature gas corrosion in aviation, rocket building, power generation, metallurgy, and engineering (Kazak and Didenko, 2016).

Ceramic coatings can be divided into three groups, namely oxide, silicon-based, and oxygen-free ceramics. *Oxide ceramics* consist of pure refractory and chemically inert oxides, mainly Al_2O_3, ZrO_2, and MgO.

Alumina (Al_2O_3) can be considered one of the most chemically inert and heat-resistant compounds. At 1705–1815 °C, Al_2O_3 is resistant to any gas, stable both in oxidizing and highly reducing atmospheres (Gevorkyan and Nerubatskyi, 2009). Thus, alumina is able to create a thermal diffusive barrier on the surface of a protected alloy. Content of alumina in a coating can often reach 80 and even 90%.

Zirconia (ZrO_2) at ca. 1000°C transforms its monolithic crystalline to a tetragonal structure (Gevorkyan et al., 2010). Since the transition is accompanied by substantial volume change of ca. 10%, zirconia can be successfully applied as a self-removable technological coating. In addition, its relatively low thermal conductivity promotes destruction of such a temporary coating.

Zircon ($ZrO_2 \cdot SiO_2$ or $ZrSiO_4$) is a silicate mineral with a strong bond between zirconia (ZrO_2) and silica (SiO_2), which makes zircon an exceptionally stable compound with low thermal expansion and thermal conductivity, excellent corrosion resistance even at temperatures as high as 1285–1700°C (Kaiser et al., 2008).

MgO ceramics have excellent thermal and mechanical properties with a very high melting point of 2800°C (Jiang et al., 2017). Magnesium oxide is hygroscopic and soluble in water, therefore, calcined MgO is used in coatings applications. Added to a coating, MgO makes it inert toward several types of steel and promotes self-removability.

Chromium oxide (Cr_2O_3) is applied for enamels as a green pigment, but in hot forming it is added as a refractory compound. In addition, it improves wettability of coatings.

Magnesium aluminate spinel ($MgAl_2O_4$ or $MgO{\cdot}Al_2O_3$) is known for its high refractoriness, low thermal expansion, chemical stability, thermal shock resistance, and corrosion resistance, which make it an attractive refractory material for the steel and cement industries (Tripathi et al., 2003).

As a rule, refractory ceramic materials, such as aluminosilicate, silica sand, or chromium-based are introduced to coatings in form of milled powders. They display chemical inertness and improve heat resistance of enamels.

In order to obtain specific properties of a protected metal surface, some non-oxygen substances are added to protective coatings, such as *metals*, *intermetallics*, *carbides*, and *nitrides*.

Among the *metals* used for protective technological coatings, various powders can be applied, such as ferroaluminum alloy, aluminum, iron, or titanium. Grain dimensions of metals are normally below 100 μm. However, metallic powders require caution and safety measures directly linked to their intensive particle oxidation. For instance, aluminum powder presents a risk due to its high explosibility, particularly when dispersed in air (Gascoin et al., 2009).

Intermetallic compounds are formed from electropositive and electronegative metals which chemically bond to form compounds with a specific composition and crystalline structure (Mattox, 2010). An ordered crystallographic structure of intermetallics is formed when the concentration of alloys exceeds the solubility limit.

Among non-oxide materials for high-temperature applications, molybdenum disilicide ($MoSi_2$) is widely used to protect Mo-based alloys from oxidation due to its excellent oxidation resistance at high temperatures and low preparation cost (Zhang et al., 2018). $MoSi_2$ is a high-melting-point intermetallic compound noted for its excellent resistance to oxidation and thermal expansion coefficient closer to that of metallic substrates comparing to other oxides. Therefore, oxide spallation from the substrate caused by a large mismatch of thermal expansion coefficient in thermal cycles can be avoided. It is thought that by further oxygen diffusion into the substrate following $MoSi_2$ formation, a thin layer of SiO_2 is formed over the $MoSi_2$ layer, which protects the underlying metal from further inward oxygen diffusion and oxidation (Seo et al., 2012).

Boron carbide (B_4C) is a hard refractory material insoluble in water and chemically inert below 700°C. Above this temperature, oxidation of B_4C occurs and forms a B_2O_3 oxygen barrier. Sun et al. (2019) demonstrate that after boron carbide is converted into B_2O_3, it gives rise to secondary coating growth on substrate particles, but as temperature continues to grow, a drastic evaporation of B_2O_3 occurs. When boron carbide is introduced to vitreous coatings, it can be applied for protection of tool steels during heat treatment in order to improve their surface hardness and adhesion between a coating and substrate.

Titanium carbide (TiC) exhibits high melting point, hardness, mechanical strength, thermal conductivity, and corrosion resistance (Grebenok et al., 2016). Moreover, it is wettable by nickel, chromium, cobalt, and iron, forming cermets with them. In turn, *silicon carbide* (SiC) is highly resistant to oxidation in a oxidating environment, since during passive oxidation it liberates SiO_2 that forms a dense layer which acts as an anti-oxidation protective coating for the surface of SiC (Roy et al., 2014).

Boron nitride (BN) is a promising interphase material with better oxidation resistance than pyrocarbon. Furthermore, after oxidation, boron and its compounds give a B_2O_3 glassy phase with a low melting point of 450°C, turning into an efficient protective layer (Willemin et al., 2017).

Additional materials can be introduced in order to achieve required performance and functionality of technological coatings. For instance, to obtain proper suspension, clay is introduced as an additive when a mixture is milled, for example, from frit, water, and chromium oxide. The suspending ability of clay depends mainly on its dispersion and is not determined by its composition, while heat resistance of coatings depends on composition and fire resistance of a clay. Its main compound is kaolinite, but bentonite clays are also used, where the basis is the mineral montmorillonite $Al_2O_3{\cdot}4SiO_2{\cdot}nH_2O$. Bentonites possess a remarkably high suspending ability and swell strongly in water.

Clay is added to most technological coatings, except for organosilicate ones containing organosilicon varnishes and organic solvents like acetone or toluene. To obtain stable suspension, water is commonly used which must be free from mechanical contaminations and impurities. In order to improve wettability, some surfactants are added during the milling or immediately after. Binders are introduced to enhance mechanical strength of coatings.

2.2.3 Composition of Coatings

With regard to chemical and physical composition of a coating, the following groups can be distinguished (Lavrentiev and Lavrentiev, 2015):

1. Glass-like coatings or enamels
2. Glass-ceramics
3. Metallic glasses
4. Organosilica
5. Protective grease
6. Metals
7. Coating systems

The above classification indicates the nature of main components responsible for protection of a metal surface during heating. Each group can be divided into subgroups dependent on proportions of the major components.

Glass-like coatings and enamels are widely used for protection of metals and alloys during hot pressing, hot rolling, and other hot working processes due to their capability of gradually turning to liquids of varying viscosity dependent on temperature. Frit composition determines most engineering properties of a coating. A glass-forming component, silica (SiO_2) or borate (B_2O_3), is the major constituent of the frit, forming two distinct glass, silica-based and borate-based series. Other oxides are usually added, e.g., Na_2O, K_2O, and Li_2Om, to control melting temperature of the frit (Rossi et al., 2021). Enamels are effective as high-temperature lubricants during hot forging. They reduce deformation resistance of a workpiece and heat transfer between a workpiece and die, thus reducing the cooling rate. As a result, forming accuracy is increased and durability of dies can be improved by 50% to even 150% (Alexenko et al., 2016).

Glass-ceramics are partially crystallized glasses composed of a combination of a crystalline phase and an amorphous glass phase (Rahaman et al., 2018). The amount of crystalline phase can vary over a wide range, from 1 to 99%, though normally in the range 30–70%. Zhao (2011) emphasizes that glass-ceramics are manufactured from base glasses, using the mechanisms of controlled nucleation and crystallization, but they exhibit special properties characteristic of both glass and ceramic materials. Consequently, special combinations of properties can be achieved and even materials with novel properties, known neither in glass nor in ceramic materials, can be designed in this group.

Metallic glasses are solid alloys that are not crystalline, having an amorphous atomic arrangement inherited directly from the liquid state (Greer, 2014). They may consist of pure metal alloys or combinations of metals and metalloids, prepared by extremely high-speed cooling techniques, such as splat cooling. These glasses may be binary (e.g., $Ni_{60}Nb_{40}$, $Cu_{50}Zr_{50}$, $Pd_{80}Si_{20}$, $Au_{73}Ge_{27}$, $Co_{75}P_{25}$, etc.), ternary (e.g., $Pd_{77}Au_5Si_{18}$, $Ni_{80}Si_8B_{12}$, $Fe_{81}Cr_2B_{17}$, $Ti_{72}Fe_{12}Si_6$, etc.), quaternary (e.g., $Fe_{40}Ni_{40}P_{14}B_6$, $Ni_{68}Cr_{10}Si_{10}B_{12}$, $Fe_{75}P_{15}C_6Al_4$, etc.), etc. systems (Karmakar, 2016). When a metallic glass is annealed close to its glass-transition temperature T_g, the glassy structure evolves toward a metastable equilibrium state of a supercooled liquid and many physical and mechanical properties change accordingly. At higher annealing temperatures, property changes are initially faster, but their total extent is more limited. The yield strain σ_y/E is much higher for MGs than for any conventional metallic materials.

Based on inherent advantages of silica-based coating, *organosilica* coatings have been designed recently to meet growing requirements (Song et al., 2018). Organosilica or organically modified silica is a molecularly inorganic–organic composite with the empirical formula of $R'_nSi(OR)_{4-n}$, that combines basic advantages of inorganic silica-based coatings and of organic groups. Generally, organosilica monomer contains a hydrolysable siloxane group $Si(OR)_{4-n}$ and reactive substituents R′, where a siloxane group can form an inorganic sol-gel coating by hydrolysis and condensation. By changing the organic R′ moiety, the coating can be endowed with antifouling function. Typically, poly(ethylene glycol) (PEG) as a representative hydrophilic polymer is reported as incorporated with a silica-based coating for the purpose of enhancing surface antifouling properties, especially protein resistance ability and enhanced coating stability (Song et al., 2018).

Various *mixtures* and *greases* spread directly on a workpiece surface before further processing constitute a distinct group of the protective temporary technological coatings. These are simplified coatings that provide significant protection, though more limited than normal coatings do (Alexenko et al., 2016).

Various *metal coatings* are used to protect components and workpieces during hot working. As a rule, metal coatings based on iron, aluminum, chromium, and copper are used. Coating of a workpiece surface is realized, e.g., by cladding with mild steel or copper, or by plasma deposition of thin layers of heat-resistant alloys. Electroplating with copper coatings affords protection of individual sections of steel shafts and gears from carburization during chemical thermal treatment. Aluminum-based coatings are formed by diffusion saturation and provide high thermal resistance.

Coating systems can be distinguished from any known type of coating because they employ two protective layers. One of them may be used during hot forming or heat treatment, while the other may remain as a protective coating during further exploitation of a component. Direct comparisons of performance of a homogeneous single-layer coating system with that of a double-layer coating system demonstrate that the latter can significantly reduce surface temperature and interfacial normal stress (Yang et al., 2009). Even though the double-layer system is more resource-consuming, it is still economically effective due to enhanced protection and subsequent improvement of final product quality.

2.2.4 Temporary Protective Coatings in Micromachining

Temporary technological coatings can also be used in microscale machining processes. Application of sacrificial protective paraffin coating is reported to substantially improve surface quality during drilling. The amount of overcutting was reportedly reduced from 4.75 to 0.56%, basically suppressing overcutting and restricting the crack initiation process (Baek et al., 2013).

Similarly, Zhao, Huang et al. (2021) publish their results on micro-ultrasonic machining (μUSM), where a slight lateral vibration at the end of a tool inevitably occurs due to tool manufacturing installation errors and mechanical system vibrations. These vibrations cause overcutting and edge breakage, creating undesirable effects on heat transfer and flow deflection. Hence, the authors present a new fabrication method that ensures high surface integrity by applying a protective coating on the substrate. PE-Wax coating of 400 μm thickness, feed rate of 4 μm/s, 60% ultrasonic power, and 25% slurry concentration proves optimum. These conditions ensure decrease of overcutting from 74.3 to 28 μm (62.3% improvement) and of weighted edge damage from 52.3 to 16 μm (69.4% improvement). Finally, a 65.85% improvement in the surface integrity index arises on application of a temporary protective coating during μUSM (Zhao, Huang et al., 2021).

2.3 HIGH TEMPERATURE COATINGS

Due to harsh work conditions, turbine blades are usually covered with coatings that display various properties like oxidation and hot corrosion resistance, strength maintenance at elevated temperatures, etc. Shourgeshty et al. (2016) divide high-temperature damage into three general groups in line with temperature ranges:

1. *High-temperature corrosion type II* (600–850°C), where sulfates are formed from the substrate at a certain partial pressure for sublimation of sulfur trioxide. The sulfate reaction with alkali metal forms low-melting-point particles that prevent formation of a protective layer.
2. *High-temperature corrosion type I* (750–950°C), includes transportation of sulfur from a deposit (sulfate base like Na_2SO_4) through an oxide layer into a metal substrate resulting in formation of stable oxides. After a reaction between a stable sulfide and sulfur moving through a scale, the base metal sulfides form a disastrous sequence in the molten phase due to high temperatures. Thus, formation of NiS_2 (molten at 645°C) and Co_xS_y (lowest liquids at 840°C) can lead to serious component degradation. The most suitable materials which can resist type I hot corrosion are $PtAl_2$-(Ni-Pt-Al) coatings (aluminide coatings modified with platinum) and MCrAlY coatings containing up to 25 wt% Cr and 6 wt% Al.
3. *Oxidation* (950°C and higher), which depends on transportation of cations or anions through the structure of an oxide layer and grain boundary. In order to form a continuous oxide layer for cobalt base superalloys, Cr content should be at least 25%. To increase oxidation resistance of chromia, addition of aluminum is preferred, especially for severe and critical conditions experienced by gas turbine blades. However, when thermal cycling conditions prevail, oxide scales can spall from a substrate surface as a result of thermally induced stresses. In the event, the oxidation resistance can be improved by addition of reactive elements to alloys and coatings, such as Y, Hf, and Ce.

Thus, the main purpose of coatings is to palliate for a poor oxidation resistance of a base alloy (aluminide, Pt–aluminide, and MCrAlY) at a high temperature. By analogy, the authors distinguish three types of high-temperature coating:

1. *Diffusional coatings. Chromium-rich coatings* are resistant toward type II high-temperature corrosion and provide some major benefits in protecting Ni-based alloys from corrosion by sulfatic deposits in chemical plants. *Aluminide coatings* usually consist of an outer layer with an aluminum-rich β-NiAl phase and an internal area rich in Ni. Aluminide coatings are fabricated through aluminizing by two different processes which differ in terms of aluminum activity in the gas phase and temperature of the process, namely the low activity–high temperature (LAHT) process aimed at development of a β-NiAl coating and the high activity–low temperature (HALT) method providing a δ-Ni_2Al_3 coating which requires subsequent heat operations to convert it to β-NiAl phase. However, aluminide coatings lose their flexibility at temperatures less than 750°C due to the internal diffusion and oxidation and subsequent loss of aluminum protection from the surface. To

address this issue, *platinum-modified nickel aluminide coatings* can be produced in two forms, namely a two-phase PtAl2-(Ni-Pt-Al) or a single-phase Pt-modified β-NiAl. Platinum supports stability of an aluminide coating in several ways, providing an aluminum-rich surface phase able to form a continuous alumina oxide shell and improve adhesion of the oxide shell formed on the surface of the coating despite thickening and increasing residual stresses. Moreover, being a refractory metal, platinum strengthens the outer layer and increases its hot corrosion resistance.

2. *Overlay coatings* are fabricated using a pre-alloyed material, such as powder, so that chemical composition of a starting material determines the final composition of coatings, unlike diffusional coatings, where composition depends on chemical composition of a substrate. The major advantages of these coatings include flexibility in choice of coating composition, increased resistance to high-temperature corrosion and oxidation in comparison with diffusional coatings, variety of coating thickness, and high ductility compared with other, especially diffusional, coatings. A typical example of an overlay coating is MCrAlY with a two-phase β+γ microstructure, where the presence of γ phase increases ductility of the coating and improves thermal fatigue resistance. 'M' stands for a combination of both Ni and Co, and Al content is typically around 10–12 wt%. Moreover, 1 wt% of yttrium (Y) is usually added to enhance adherence of the oxide layer.
3. *Thermal barrier coatings* (TBCs) are widely used for critical high temperature conditions that take place in combustion chambers or rotating blades. Thermal barrier-coated components must withstand the most extreme temperature, temperature cycling, and high stress. They are expected to withstand thousands of takeoffs and landings in commercial jet engines and up to 30,000 h of operation in industrial gas turbine engines. TBCs exhibit important advantages, namely increasing lifetime of parts, improving engine efficiency since it allows for increasing turbine inlet temperature, and decreasing coolant air flow. These are achieved through combination of the multi-material nature of the TBC structure, which in addition to the demanding operating conditions makes TBCs more complex than any other coating system. Typically, a TBC coating system consists of four layers totally different from each other, namely:
    1. Super alloy substrate
    2. Aluminum intermediate bond coating
    3. TGO (thermally grown oxide)
    4. Ceramic final outer coating

The term "nano-coatings" sometimes refers to TBCs that are kinds of thin layers which are in nano dimensions or have substrates in which nanoscale particles have been dispersed and produce special properties. In many cases, their properties show significant improvements, in particular, have a higher coefficient of thermal expansion, hardness, and toughness as well as higher resistance to corrosion, abrasion, and erosion in comparison to micrometer-scale coating structures.

Four important groups of nano-coatings can be distinguished (Shourgeshty et al., 2016):

1. Nano-grade coatings
2. Superlattice and multilayer coatings
3. Thin-film coatings
4. Nanocomposite coatings (Shourgeshty et al., 2016)

A distinctive family of ceramic materials has come to be known as *ultra-high-temperature ceramics* (UHTCs). It includes ceramic borides, carbides and nitrides of early transition metals such as Zr, Hf, Nb, and Ta, especially hafnium diboride and zirconium diboride-based compositions, characterized by high melting points, chemical inertness and relatively good oxidation resistance in extreme environments (Gasch et al., 2005). Fahrenholtz and Hilmas (2017) point out that UHTCs are often defined as compounds that have melting points above 3000°C or, in the most pragmatic way, calling UHTCs ceramic materials that can be used for extended times at temperatures above 1650°C. However, none of these definitions captures the wide range of extreme conditions in which UHTCs may be used. The strong covalent bonds between the transition metals and B, C, or N produce UHTCs not only with melting temperature, but also with high hardness and stiffness, as well as higher electrical and thermal conductivities than that of oxide ceramics. The authors emphasize that the above-mentioned intriguing combination of metal-like and ceramic-like properties allows UHTCs to survive extreme temperatures, heat fluxes, radiation levels, mechanical loads, chemical reactivities, and other conditions that are beyond the capabilities of existing structural materials (Fahrenholtz and Hilmas, 2017).

Gnanavelbabu et al. (2021) note that by adding ceramic reinforcements, it is possible to enhance corrosion resistance of a magnesium alloy and concentrate on UHTC reinforcements. They analyze Mg-AZ91D matrix composites reinforced with 5 wt% of TaC, HfC, TiN, TiB2, and TiC particles. The authors demonstrate that the addition of reinforcing UHTC particles increases density and decreases porosity of the obtained composites, as well as raises hardness of the composites as compared to pure Mg-AZ91D material. They also find that higher corrosion resistance could be obtained due to formation of a dense and stable anti-corrosion protective layer and reduction of β-Mg17Al12 phase.

Considering some aspects of heat transfer in interactions between components with a sharp leading edge and high-enthalpy high-speed flows of dissociated air, Simonenko et al. (2013) name some characteristics of the UHTC materials which would make them promising for use in hypersonic flight vehicles:

1. Besides high melting points and phase stability in a wide temperature range, the materials must have comparatively high oxidation resistance, in particular, in reactions with atomic oxygen.
2. The material should have a high thermal conductivity that must ensure heat removal from strongly overheated regions, which is especially important for samples with a sharp leading edge, since local overheatings to ~2500°C

cause softening and ablation and lead to deformation of geometry of components affecting flight performance.

3. The material must have minimum catalytic activity in exothermic reactions of surface recombination in order to minimize the chemical component of aerodynamic heating.
4. It must exhibit a high emissivity coefficient and high thermal conductivity, which provides transfer of energy inside the system, and, as a result, from the system to the outside environment.

Morks et al. (2013) describe an ultra-high-temperature ceramic (UHTC) as a potential candidate material for advanced aerospace vehicles. Thermal protection systems in rocket exhaust cones, insulating tiles for space shuttles, engine components, and ceramic coatings embedded into windshield glass of many airplanes consist of multilayer ceramic coatings that incorporate mullite, yttria stabilized zirconia (YSZ), and ultra-high-temperature zirconium diboride (ZrB_2). In advanced aircraft, it is customary to represent the speed of aircrafts in supersonic terms, exceeding the speed of sound (Mach 1), and hypersonic terms, five times and above the speed of sound. In the conditions of supersonic speed, jet engine components are compressed and thrust by a huge amount of hot gases at above 2500°C. Increasingly severe operating environment for high-temperature structure materials requires more reliable TBCs to improve performance of structural materials and extend the operating time and temperature in these extremely harsh conditions.

According to Morks et al. (2013), the mullite layer in TBC reaches its melting point within a short time at a high operating temperature (>2500°C). The molten mullite may seal the stressed ZrB_2 layer penetrating into cracks and pores of ZrB_2, which sheds some light on what could happen at extreme temperatures. Namely, the mullite layer between ZrB_2 and YSZ layers is to repair the ZrB2 layer that suffers from thermal stresses providing ultra-high-temperature protection. In fact, the self-sealing for the cracks in ZrB2 layer with molten mullite is the most important feature of this UHTC multilayer material (Morks et al., 2013). What is more, self-healing and self-repair is one of the most important directions in development of components.

Fahrenholtz and Hilmas (2017) note that the family of UHTC materials being studied at present essentially consists of the same materials that were examined during the space race of the 1950s and 1960s. In their view, discovery of new materials is needed to expand the number of UHTCs and to extend the number of their potential applications. They believe that the likelihood that a large number of new UHTC compounds remain undiscovered is extremely low, which may indicate a direction for development of advanced structural UHTC materials.

2.4 THERMAL SPRAY COATINGS

In order to restore dimensions of worn or corroded components, thermal spray (TS) coatings are used in every manufacturing industry. Dorfman (2018) underlines that although the thermal spray coating does not add any strength to an existing material, it is a quick and economical way to restore dimensions of components. After a part is coated, subsequent grinding operations are often needed to smooth its surface and to bring the final dimensions to required tolerances. Li (2010) notes that apart from dimensional restoration, thermal spray coatings can add some functions to light metals, such as wear resistance, corrosion resistance, bioactivity, and dielectric properties. Some characteristics of the deposition process, such as splat cooling and successive stacking of splats, lead to creation of coatings with unique microstructures different from conventional materials.

A typical thermal spray system consists of five subsystems (Fauchais et al., 2014):

1. A high-energy, high-velocity jet generation unit, including a torch, power supply, gas supply, and associated control systems
2. Coating material preparation and its transformation into a stream of molten droplets, including the powder size distribution and morphology, and its injection into a high-energy gas stream
3. Surrounding atmosphere and its controlled parameters including humidity, low pressure, etc.
4. A substrate material with a properly prepared surface
5. Mechanical equipment for controlling motion of the torch and the substrate relative to each other, i.e., stand-off distance, velocity, etc.

According to Dorfman (2018), there are some key decisions to be made for salvage and repair materials, including coating thickness, color match, surface profile after finishing, ease of finishing, bond strength, deposition rate, and application cost. As an example of a salvage and repair material, he mentions nickel 5 wt% aluminum powder, manufactured by a variety of different processes, that may be mechanically clad, chemically clad, or gas- or water-atomized. The choice of a process is based on the application, considering that similar chemical compositions do not always guarantee equivalent performance. For instance, Ni 5 wt% Al coatings with higher oxide levels are harder and more difficult to machine than coatings with lower oxide levels. If a coating exhibits poor cohesive strength and unmelted particles appear in it, this may result in particle pullout during finishing and finally in a porous surface finish. In addition, a coating with unreacted aluminum within its microstructure may be a problem for parts exposed to the environment of corrosive substances (Dorfman, 2018). Characteristics of thermal spray processes have impact on spray particle parameters and metal alloy particle oxidation; splat formation (including solid–liquid two-phase droplet impact involved in cermet coating deposition) influence features of pores, coating microstructure, dominant effect of lamellar structure on coating properties, and reactions of spray particles with light metal substrates (Li, 2010).

Łatka et al. (2020) describe *thermal spray techniques* used frequently to obtain functionally graded coatings, namely atmospheric plasma spraying (APS), suspension plasma spraying (SPS), solution precursor plasma spraying, high-velocity oxy-fuel and high-velocity air-fuel spraying, low-pressure cold spraying, and high-pressure cold spraying.

Atmospheric plasma spraying (APS) is perhaps one of the most frequently used methods of coating manufacturing in the field of TS. A detailed description of this method can be found in Heimann (2008) or Pawłowski (2008). During processing, an electrical arc is ignited between a copper anode and a thoriated tungsten cathode ionizing plasma gases, heating and expanding them to form a plasma jet. The powder particles are transported by a carrier gas and injected into the hot stream of plasma jet, where they are heated and accelerated by drag force. Directed to the substrate surface, these particles hit it with a relatively high kinetic energy and form splats that solidify and build up the coating. Commercially available low-power plasma torches feature powers up to 80 kW, but typical torches can operate with an electric power input of up to 200 kW, ensuring a powder feed rate of up to 100 kg/h.

The microstructure of sprayed coatings results from the phenomena occurring inside a powder particle in plasma jet dependent on various process parameters. Composition of a working gas is the fundamental process parameter, where two main groups of gases are used: (1) primary gases to stabilize arc inside the torch nozzle, frequently argon Ar and sometimes nitrogen N_2; and (2) secondary gases added in order to increase heat conductivity of plasma. Other important APS parameters listed by Fauchais et al. (2014) include electrical power, flow rate of the plasma gases, feed rate of the powder, particle size distribution of the powder (usually between 20 and 90 μ), and speed of plasma torch relative to the substrate. According to Rico et al. (2018), power is the key parameter governing the microstructure and mechanical performance of the coating, but other parameters, such as the arc current, plasma gas flow rate, carrier gas flow rate, or stand-off distance can also have effect and thus should undergo optimization.

A proper APS process leads to melting of injected particles before their impact on the substrate through an appropriate choice of spray parameters. However, the powders usually include small and large particles, which somewhat disrupts the process and affect coating properties. On the one hand, small particles may get molten and start to evaporate in plasma, weakening the coating effect, on the other hand, large ones may remain solid and form weakening inclusions in a formed structure. Generally, the powder is injected radially, with injection angles ranging from 75° to 120° relative to the torch axis, with injectors placed outside the plasma torch, though some torches are designed to feature injection inside the torch (Łatka et al., 2020).

APS is the preferred method for spraying high-temperature ceramic oxides and the large number of adjustable process parameters allows adaptation of the coating process to a wide range of materials and substrates with different properties (Fauchais et al., 2014). However, the excessive number of process parameters (up to 60) also becomes a disadvantage since proper control of all these parameters requires special effort. According to Pawłowski (2008), a typical adhesion strength of APS-obtained coatings is in the range from 15 up to 30 MPa, but application of a bond coat, such as Ni-Al or Ni-Cr or Mo, may increase the adhesion strength up to 70 MPa. APS coatings are typically a few hundreds of micrometers thick.

Suspension plasma spraying (SPS) is able to produce submicron- and nano-grained coatings due to application of suspensions instead of dry powders as starting materials. SPS can be considered an advancement in APS as it enables spraying of fine feedstock of 100 nm–5 μm diameter (Mahade et al., 2019). Microstructural features inherently associated with the process, such as a smaller splat size, fine scale porosity, or low surface roughness enhance wear performance of SPS coatings. Apart from thermal barrier coatings with improved performance compared to APS coatings (Kassner et al., 2008; Curry et al., 2014; Bernard et al., 2017), there are reports on deposition of nanostructured WC–12% Co coatings by the SPS method (Berghaus et al., 2006), submicron silicon carbide composite coatings avoiding the decomposition of SiC typically expected with APS (Mubarok and Espallargas, 2015), as well as pure titanium and chromium carbides wear-resistance coatings (Mahade et al., 2019). SPS is advantageous for deposition of a functionally graded double ceramic insulation layer of $La_2Zr_2O_7$/8YSZ (Wang et al., 2015) or for multilayered functionally graded coatings with a compact microstructure exhibiting limited presence of pores and other defects, which results in good mechanical properties (Cattini et al., 2013). It should be added that porosity can vary dependent on different parameters from 5 to 11.6% for YCeSZ powders and from 10 to 27% for YSZ powders (Sokołowski et al., 2017), while SPS with a < 3 μm powder suspension has generated the most porous coating of 41 $\pm$ 2% open porosity, exhibiting the smallest thermal conductivity of 0.56 $\pm$ 0.1 W/m·K (Ganvir et al., 2015).

The SPS process offers various possibilities to design microstructure of coatings, for instance (Łatka et al., 2020):

- Dimensions of pores may vary from nanometer through submicrometer to micrometer size, having various shapes and forms connected or non-connected networks.
- Morphology may vary from dense and homogeneous through vertically cracked to fully columnar.
- Deposits may be formed by fully molten splats, sintered particles only, or by both (the so-called two-zone structures).
- Roughness of obtained resultant coated surface may vary in a very broad range with *Ra* values between 1.6 and 14.1 μm.
- Thickness of a coating may range from few micrometers to hundreds of micrometers.

Thermal spraying with a *hybrid powder-suspension* feedstock is proposed as a novel approach to obtain coatings with unusual chemistries and unique microstructures. In particular, Ganvir et al. (2021) explore benefits of a hybrid powder-suspension system depositing Al_2O_3-YSZ ceramic matrix composite (CMC) coatings with unique multi-length scale microstructures for tribological applications. They demonstrate that as-deposited CMC coating exhibits significantly improved tribological properties by enhancing wear resistance under scratch, dry sliding ball-on-plate (ca. 36% decrease), and erosion tests (50% decrease in the erosion wear rate) as compared to the conventional APS deposited monolithic Al_2O_3 coating.

Similarly, Mahade et al. (2020) describe coatings deposited by simultaneous spraying of T-400 (Tribaloy-400) powder and Cr_3C_2 suspension. The as-obtained coating revealed a lamellar microstructure with distributed fine carbides, showing the presence of original feedstock constituents, along with some oxides of chromium. Hardness of the as-sprayed coatings is higher than in the pure T-400 coating. Moreover, the hybrid coatings possess excellent wear and scratch resistance, superior compared to the pure T-400 coating, which is important for durability improvement of engineering components operating under severe wear conditions. The authors believe that the hybrid approach is easily extendable to other material systems and can contribute to fabrication of next-generation wear-resistant coatings.

Solution precursor plasma spraying (SPPS) allows for manufacturing coatings with submicrometer and nanometer structures. Examples of the coatings obtained with SPPS include thermal barrier coatings (Yang et al., 2020), nanostructural photocatalytic coatings (Dom et al., 2012), hydrophobous rare earth coatings (Xu et al., 2021), coatings for solid oxy-fuel cells (Shri Prakash et al., 2017), environmental barrier coatings (Darthout et al., 2016), bioactive glass coatings (Canas et al., 2019; Garrido et al., 2021), coatings for gas sensors (Yu et al., 2017), etc. The precursors applied to spraying are generally mixtures of compounds, such as salts, acetates, or nitrates, dissolved in a solvent (usually an organic liquid or water) to form a final solution (Łatka et al., 2020).

There are also reports on *hybridizing the conventional atmospheric plasma spraying (APS) technique with the solution precursor plasma spray (SPPS)* route. This way, Joshi et al. (2014) produce thermal barrier coatings (TBCs) with tailored configurations. The authors claim that such a hybrid process can be conveniently adopted for forming composite, multilayered and graded coatings, yielding distinct TBC microstructures with extended coating durability. Comparing TBC specimens generated using the conventional APS technique, the SPPS method, and APS-SPPS hybrid processing, the authors demonstrate advantages of the hybrid processing. A similar hybrid APS-SPPS route with simultaneous feeding of an appropriate solution precursor and commercially available spray-grade powder feedstock is reported to enable fabrication of microstructures with nanostructured and micron-sized features (Lohia et al., 2014). Hou et al. (2019) propose a hybrid suspension-solution precursor plasma spray process with a radio-frequency thermal plasma torch and demonstrate it is feasible to deposit $Ba(Mg_{1/3}Ta_{2/3})O_3$ (BMT) nanostructured coatings.

In *high velocity oxy-fuel* (HVOF) and *high velocity air-fuel spraying* (HVAF), gas combustion serves as an energy source for melting and accelerating powder particles. The HVOF process enables reaching higher particle velocity and lower particle temperature than APS and thus high deposition efficiency, good adhesion of coatings with a low content of oxides and low porosity (Łatka et al., 2020). Functionally graded coatings (FGCs) with gradually changing coefficient of thermal expansion and Young's modulus, which decreases stress and increases bond strength, may be obtained by HVOF in two ways (Mamun and Stokes, 2014):

1. Using premixed powders
2. Co-injection of different powders

Łatka et al. (2020) underline that coating quality can be controlled by particle in-flight temperature, while the most important parameters influencing the deposition of FGC are spray distance and flow rate of working gases, fuel, and oxygen. Moreover, they suggest paying attention to control of the powder with the lowest melting point in order to avoid excessive evaporation.

Low-pressure cold spraying (LPCS) and *High-pressure cold spraying* (HPCS) processes both utilize the main advantage of cold spraying, namely, the low processing temperature compared to other thermal spray processes. Cold gas dynamic spray is a solid-state process, based on the kinetic energy of incident particles to deposit coatings (Sabard and Hussain, 2019). In the cold spray processes, metallic powders are accelerated by a supersonic gas flowing through a convergent–diverging nozzle. Impact of particles on a surface causes combined plastic deformation of the feedstock material and the substrate and thus formation of a coating. Sprayed materials are kept below melting temperature, so that any deleterious effect of the temperature during deposition is avoided. Most probably, the accumulation of dislocations during severe plastic deformation is responsible for various grain structure formations of cold-sprayed deposits in the interfacial regions, from elongated grains between 200 nm and 1 μm to ultrafine grains (50–200 nm) (Sabard and Hussain, 2019).

Łatka et al. (2020) categorize cold spraying methods according to the initial pressure of working gas:

1. High-pressure cold spraying at a gas pressure above 1 MPa, usually with nitrogen or helium as the working gas
2. Low-pressure cold spraying at a gas pressure below 1 MPa with nitrogen or air

In both the methods, the working gas is heated before reaching the nozzle. The LPCS equipment and processing costs are much lower than in HPCS, but since the sprayed particles reach much lower velocity in the former, its application is significantly limited to deposition of easily plastic deformable materials such as tin, zinc, copper, aluminum, nickel, and some composite materials. Especially metal matrix composites (MMCs) can be fabricated for various purposes, such as regeneration of surfaces, increase of electrical conductivity, corrosion and wear resistance improvement, biomedical applications, etc. (Łatka et al., 2020).

Lazorenko et al. (2021) in their review emphasize that thermal spray coating techniques have been developed recently to protect rail steel from atmospheric corrosion, mainly due to its simplicity, relatively low cost, and high efficiency. They give an example of effects of atmospheric corrosion in a coastal zone on carbon steel with Zn, Al, and Zn-Al thermally sprayed coating after 33 years, where corrosion of steel with a zinc coating was effectively slowed down by zinc corrosion products formed on the surface, while long-term corrosion resistance of aluminum coatings was facilitated by a thin surface oxide film. They provide similar examples of good condition of 40- and 50-year-old bridges with thermal spray zinc duplex coatings, stressing that lower durability of the coatings (20 years) was caused by some errors, such as pinholes, spitting, and low coating thickness.

Pathak and Saha (2017) indicate that restoration of dimensionally inaccurate components or especially damaged parts through cold spray has become a great achievement, especially in reparation of surface damage to intricate shapes. In the process of restoration, a metallic powder is deposited in distorted sections. The host and powder materials may be dissimilar, which is very useful to improve the various mechanical properties of a part, depending on requirements. Unlike other repairing processes, cold spray does not generate undesirable thermal stresses that can result in premature failure of a component. The authors emphasize that this process is being widely used in the automotive and industrial marketplaces and is ideally suitable for aerospace materials like magnesium and titanium. They list benefits and examples of repair and remanufacturing by cold spray:

- Significant total cost savings.
- Repair time reduction. Cold spray can be used to repair a part in-situ and eliminate assembly and testing costs.
- Improved production yield. Cold spray can be applied to used-up parts returned from the field, but it can also be used to salvage parts with manufacturing defects, making them useful.

Fauchais et al. (2014) describe many areas where thermal spray coatings may be used as a repairing technique. In the case of aluminum components that are difficult to repair by welding due to its high-specific thermal conductivity and high coefficient of thermal expansion, they highlight the ability of the low-pressure cold spray technique to replace welding in repairing cracks.

2.5 VACUUM PLASMA SPRAYING

Vacuum plasma spraying refers to transport of particles from a gaseous source to a substrate surface along a linear motion trajectory (Gushcha, 2009). When vacuum is around 10^{-2} Pa and less, the process is classified as a vacuum evaporation, pressure 1 Pa constitutes the so-called cathodic spraying, while 10^{-2}–10^{-1} Pa is typical for magnetron sputtering and ion–plasma beam sputtering. Vacuum spraying is used for sputtering of metals such as Al, Au, Cu, Cr, Ni, V, Ti, alloys, e.g., NiCr or CrNiSi, chemical compounds like silicides, oxidizes, borides, carbides, etc., complex glasses, and cermets (Gushcha, 2009). Vaßen et al. (2018) state that VPS is the most often used thermal spray process for deposition of dense, high-quality 3-mm thick tungsten coatings. Kim et al. (2020) report deposition of hafnium carbide (HfC) and titanium carbide (TiC) with the VPS method to form ultra-high-temperature ceramic (UHTC) coatings in the form of both single-layer HfC and TiC and multilayer HfC/TiC coatings. Thicknesses of the HfC and TiC single-layer coatings were 165 and 140 μm, respectively, while the thicknesses of the HfC and TiC layers in the HfC/TiC multilayer coating were 40 and 50 μm, respectively. No oxides were observed in any of the coating layers and average porosity was approximately 16.8% for the HfC coating and 22.5% for the TiC coating. The hardness values of the HfC and TiC layers in the multilayer sample were 1563.5 and 1059.2 HV, respectively. Kim et al.

(2021) fabricated metallic glass coating layers out of new $Fe_{46.8}$-$Mo_{30.6}$-$Cr_{16.6}$-$C_{4.3}$-$B_{1.7}$ powders via VPS. Despite presence of splat particles and minor unmelted powders, all the alloying elements in the VPS material were evenly distributed throughout the coating layer and showed superior corrosion resistance compared to the similar material fabricated with APS method. Unlike APS, the VPS process uses a low-pressure argon atmosphere that inhibits oxidation of the coating layer, while the high velocity of the molten droplets facilitates formation of highly dense coatings (Kim et al., 2020). Henao et al. (2019) describe application of VPS to produce bioactive glass coatings on Ti-6Al-4V substrates. Deposition of high-quality coatings with good adherence and with very little or even no oxidation, high homogeneity, and without any modification of the initial bioactive glass composition was possible.

VPS systems constitute separate devices of complex functions which are able to (Romanov, 2010):

- Generate proper vacuum
- Evaporate or spray a coating material
- Control sputtering parameters and thus properties of obtained films
- Transport components before and after processing
- Ensure proper power supply

From the design perspective, a typical VPS unit consists of the following subunits (Gushcha, 2009):

- Spray chamber, where the material is sputtered on the substrate
- Sources of evaporated or sprayed materials with power supply and control systems
- Pumping and pressure-distribution systems ensuring proper vacuum and gas flow, including pumps, valves, traps, flanges and covers, as well as measurement devices for vacuum and gas flow rates
- Main power supply and safety switches for all the subunits
- Control systems for sputtering dosing including sensors, actuators, and controllers, so that proper sputtering rate and temperatures ensure desired thickness and properties of the coating
- Transporting devices to place components in the spray chamber with exact positioning, as well as to change their position during the process when a multilayer coating is produced
- Additional equipment such as screens and dampers, manipulators, hydraulic and pneumatic drives, gas cleaning devices, etc.

For practical realization of the plasma spraying process, integrated technological systems are developed, enabling both modeling and experimental study (Kundas and Ilyuschenko, 2002). The proposed system provided simulation of the main plasma spraying process stages, including heating and acceleration of particles, heat transfer, coating structure formation, and stress–strain state in a coating–substrate system, but also experimental measurement of the main characteristics of the process and computer control of the spraying process to obtain coating of the desired characteristics.

2.6 PROCESSES OF HARD ALLOYS COATINGS

Hard coatings with a thickness of a few microns are widely applied to improve performance of tools, for example, in cutting, forming and casting applications, and components, for example, in automotive and aerospace applications (Mitterer, 2014). Different methods have been developed to fabricate hard coatings, including thermally activated chemical vapor deposition (CVD), plasma-assisted physical vapor deposition (PVD), plasma-assisted chemical vapor deposition (PACVD), and laser-assisted methods such as pulsed laser deposition. The coatings synthesized with these methods offer high hardness, oxidation and wear resistance, thermal stability, diffusion barrier properties, and in some cases also low friction and even smart, self-adaptive properties (Mitterer, 2014).

Chemical vapor deposition (CVD) is a well-known chemical method where a film is deposited on the substrate surface through chemical reaction from a gas-phase or vapor-phase precursor through application of activation energy (Madhuri, 2020). Several gases, distinguished as a source gas and a carrier gas, are admitted into the vacuum chamber, and after dissociation between the species, the newly formed molecules are deposited on the heated substrate as shown in Figure 2.8.

Madhuri (2020) lists various modifications of CVD, such as atmospheric pressure CVD, low-pressure CVD, where the pressure is below ambient (0.6–1.33 mbar), laser CVD where laser is used to enhance chemical reaction, plasma-enhanced CVD where plasma enhances the decomposition process, photochemical vapor deposition, chemical vapor infiltration, and chemical beam epitaxy. The operating parameters, modes, and conditions can be selected based on properties of a substrate material and designed application. Mechanical and thermal properties and phase stability such as melting point and thermal conductivity of a coating material should be considered when choosing appropriate process parameters (Madhuri, 2020)

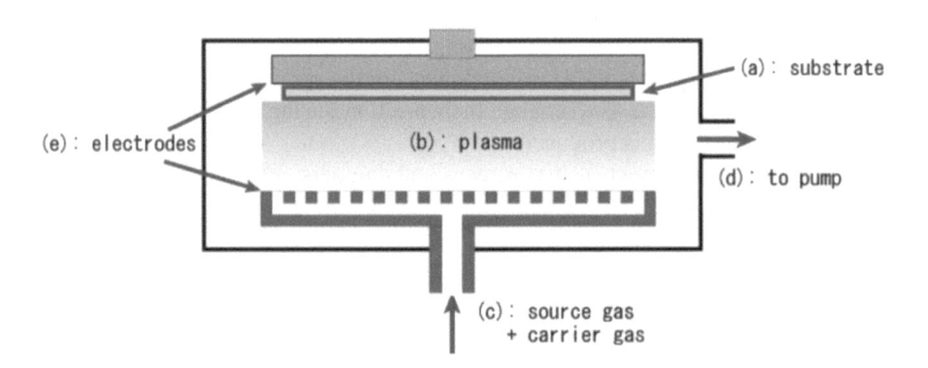

FIGURE 2.8 Schematic diagram of a plasma CVD (chemical vapor deposition) system (© Public domain).

Benelmekki and Erbe (2019) point out that CVD processes are well known and developed and reactors used for the process variations depend basically on the type of precursors, deposition conditions, and forms of energy introduced to the system to activate the desired chemical reaction. Apart from the previously mentioned modifications, the authors consider two more processes as the most established ones:

- Metal-organic CVD process which is applied when metal-organics are used as precursors, such as trimethylaluminum, $Al_2(CH_3)_6$, or trimethylgallium, $Ga(CH_3)_3$
- Aerosol-assisted CVD where liquid precursors are introduced to the CVD reactor in an aerosol form, facilitating its evaporation (Benelmekki and Erbe, 2019)

Physical vapor deposition (PVD) is a vacuum process where material is transferred in the form of vapor particles from a material source to the substrate surface where it is condensed as a film (Bouzakis and Michailidis, 2018). PVD encompasses a broad family of vacuum coating processes employing material removal from a source by evaporation or sputtering. Chemical compounds are deposited either by using a similar source material or by introducing reactive gases (nitrogen, oxygen, or simple hydrocarbons) containing appropriate surfactants and reacting with metal from the PVD source. Most PVD processes are typically named after the means used for producing the physical vapor, mainly evaporation and sputtering.

- *Evaporation* can be resistive, inductive, electron beam, activated reactive evaporation, or arc evaporation under DC or AC.
- *Sputtering* can be diode or triode, ion beam, or magnetron sputtering, i.e., direct current (DC), radio frequency (RF), pulsed cathode, dual magnetron sputtering, or high-power pulsed magnetron sputtering (HPPMS or HIPIMS: high-power impulse magnetron sputtering). A promising technology for deposition of thick films is the high-speed physical vapor deposition (HS-PVD) process based on a hollow cathode glow discharge (Bouzakis and Michailidis, 2018). Figure 2.9 illustrates the plasma spray-physical vapor deposition process.

The PVD process is able to produce a metal vapor which is then deposited on electrically conductive materials as a thin, highly adhered pure metal or alloy coating (Shang and Zeng, 2013). The process is carried out at high vacuum using a cathodic arc source and the coating is deposited over the entire object surface uniformly rather than in localized areas. The authors state that all reactive PVD hard coating processes combine three subsystems:

1. A method for depositing the metal
2. Combination with an active gas, such as nitrogen, oxygen, or methane
3. Plasma bombardment of the substrate to ensure a dense, hard coating

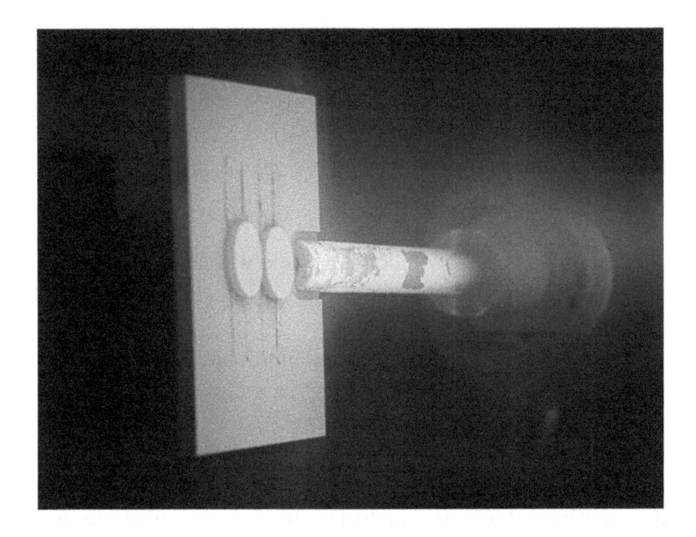

FIGURE 2.9 Plasma spray-physical vapor deposition (© Public domain).

Ion plating, ion implantation, sputtering, and laser surface alloying are among the primary PVD methods (Shang and Zeng, 2013). Ion implantation of various elements on cutting tool surfaces improves wear resistance (Morozow et al., 2018) and significantly extends their life cycle (Morozow et al., 2021). Ceramic coatings can be deposited with PVD techniques at relatively low temperatures, avoiding microstructural defects within the substrate material, such as embrittlement effects, non-uniform density, micro cracks and porosity, and weakened adhesive strength (Miorin et al., 2019).

Working pressure of PVD processes can vary from 5 to 0.1 Pa, enabling fabrication of coatings with characteristics not yet achievable with any of the existing thermal spray processes. Obtained PVD coatings are very homogeneous, dense, hard, thin, gas tight, and can have specially designed microstructures (von Niessen and Gindrat, 2011). High investment costs and low deposition rates are among the disadvantages of conventional PVD processes.

Another method of hard coating deposition is associated primarily with restoration of car parts, namely the technology of *electrodeposition of composite galvanic coatings* (CGC) (Semenikhin et al., 2020). The essence of the CGC method is precipitation of a metal from the galvanic bath, together with various powders deposited on a part, such as metals, oxides, carbides, borides, and sulfides, as well as polymers. Inclusion of dispersed materials in the metal substrate significantly changes their properties, increasing anti-friction characteristics, wear resistance, thermal and corrosion resistance, and widening application of the CGC to restoration of car parts. Among advantages of the method, comparative ease of coating disposal directly on the part surface, the low cost, the possibility of the process automation, and only slight influence of coatings on material properties of parts can be mentioned.

Semenikhin et al. (2020) emphasize that CGC can be realized in various ways, most often from a galvanic bath. The simplest version comprises an electrolyte poured into the bath and mixed with powder, a detail fixed on the cathode, and the dispersed phase maintained in a suspended state or transported to the cathode. Electric current passing through the suspension forms a coating on the surface of a restored component. According to the authors, CGC can solve three main problems:

1. Application of metal coating to a worn surface during recovery of parts and their hardening
2. Application of metal and other coatings to protect part surfaces from corrosion
3. Application of protective and decorative coatings

In terms of reliability and durability of repaired vehicle parts, the first task is the most important. Analysis of car parts coming to reparation shows that wear is mostly below 0.1–0.3 mm, which makes coating more economically effective than replacement of a worn part, especially when the parts are made of expensive alloy steels. Application of powders based on tungsten carbide with a particle size below 1 μm, produced by electro-erosion dispersion from waste-sintered hard alloys as a dispersed phase of composite galvanic coatings based on iron during restoration and hardening of car parts, is possible. The resulting micro-hardness was close to that of the original component, while relative wear resistance of the obtained composite galvanic coatings was increased compared to simple iron galvanic coatings (Semenikhin et al., 2020). It is noteworthy that not only waste materials are reused and worn parts repaired at reduced costs, but also life of repaired parts is increased.

Mayrhofer et al. (2005) point out that nanostructures have attracted increasing interest in modern development of hard coatings for wear-resistant applications. There is a direct relation between hardness and nanostructure, because hardness of a material is determined by resistance to bond distortion and dislocation formation and motion, which in turn depend on the amount and constitution of obstacles inside the material structure. Based on $TiB_{2.4}$, TiN–TiB_2, $Ti_{0.34}Al_{0.66}N$, and Ti(N,B) as model-coatings, the authors demonstrate development of self-organized nanostructures and their influence on mechanical properties of ceramic thin films. Growth by segregation-driven processes is the predominant self-organizing mechanism for two-dimensional TiB2.4 and three-dimensional TiN–TiB2 nanostructures. On the other hand, growth of $Ti_{0.34}Al_{0.66}N$ and Ti(N,B) is the result of a supersaturated TiN-based phase formation, which tends to decompose into its stable constituents via formation of nm-sized domains during post-deposition annealing (Mayrhofer et al., 2006).

Fox-Rabinovich (2013) describes application of adaptive coatings that enable a tribo-system to shift to a milder wear mode due to formation of protective/lubricious tribo-films during operation as a result of interaction with the external environment. Cutting and stamping tools can be considered as heavy loaded tribo-systems (HLTS), working under severe conditions of high temperature and stresses with intensive wear rates. Under severe frictional conditions, irreversible thermodynamic and self-organization phenomena take place that may be used as an efficient protection of a friction surface. Hard plasma vapor deposited coatings are best suited for this application. The self-organizing phenomenon is characterized by formation of thin tribo-films, from several nanometers to a micron thick, on the friction surface by interaction with the environment. Adaptive coating application weakens internal impact and leads to practical improvements in two domains: (1) tool life enhancement, and (2) improved workpiece quality in terms of a better surface finish, improved dimensional accuracy, etc. The author gives an example of Al-rich TiAlN and AlTiCrN families of nanocrystalline hard plasma vapor deposited coatings which due to their adaptability mostly outperform other categories of coatings under aggressive and extreme cutting conditions. The author is of the opinion that future generations of hard coatings used in applications under extreme tribological conditions must become complex engineering adaptive systems to be able to sustain a number of external impacts and adapt to changing operating conditions (Fox-Rabinovich, 2013).

2.7 TITANIUM NITRIDE NANOCERAMIC MATRIX COMPOSITE COATINGS

The processes and equipment for synthesis of TiN coatings are well established and commonly applied in industry. These processes are envinronmentally clean and the coatings are chemically inert toward the environment. Moreover, their application as strengthening and decorative coatings is economically profitable. Practical applications of titanium nitride involve several industrial branches and motivate new directions of physical and technical research.

Titanium nitride (TiN) is a gold-yellow ceramic material of high hardness up to 2000 kg/mm^2, high decomposition temperature of 2949°C, and high chemical stability at room temperature (van Hove et al., 2015). It has a cubic structure similar to NaCl (Tominaga et al., 2012). Along with chromium nitride (CrN), alumina (Al_2O_3), titanium carbide (TiC), diamond-like carbon (DLC), and tungsten carbide/carbon (WC/C), titanium nitride belongs to the group of important ceramic coating materials that improve toughness, tribological properties, wear resistance, and high-temperature stability of components (Maitra, 2014). Nanoceramic composite coatings consist of a homogeneous isotropic multiphase mixture or several layers of ceramic phases. Multilayer coatings have significant impact on wear mechanism, having transition zones between their layers and thus hindering crack propagation, since the cracking direction is from the coating surface to the substrate, unlike two-directional cracking in single-layer coatings (Pogrebnjak et al., 2019).

The nanocomposites are beneficial because of improved hardness and oxidation resistance higher than in coventional TiN coatings. However, nanocomposite coatings suffer from some limitations to possible particle sizes and material compositions due to restricted conditions and altered nanoscale material properties. As a rule, nanocomposite coatings are deposited using physical vapor deposition (PVD) or plasma-assisted chemical vapor deposition (PACVD). With application of laser-assisted deposition techniques, it is possible to fabricate hard composite coatings with nanocrystalline and amorphous phases. When the particle sizes are selected appropriately, optimal dislocations and cracks in nano- and microscale in a material can be achieved. This way, the material can gain self-adjustment properties under deformation beyond the elastic limit, changing it from hard elastic to plastic and avoiding brittle destruction of the material and improving its wear resistance (Maitra, 2014).

Tominaga et al. (2012) report preparation of TiN thin films by sputtering in Ar atmosphere, obtaining a crystalline structure even when the substrate temperature is low. Direct sputtering of mixed TiN and AlN powders in Ar enabled production of Ti–Al–N thin films of high mechanical strength. These films consist of crystalline TiN and amorphous AlN when the proportion of two compounds is 50/50 (Tominaga et al., 2012). It can serve as an example of the concept of second-generation Ti(X) N coatings, where X denotes a metallic element introduced to the lattice of titanium nitride to obtain more excellent resistance to wear and oxidation than of the first-generation TiN layer (Shen et al., 2021).

Research indicates that binary metal nitrides exhibit better characteristics than conventional metals do (Pogrebnjak et al., 2019). The multilayer nitride coatings find their applications in the aerospace industry, medicine, and manufacturing. Among important applications of TiN-based coatings, Santecchia et al. (2015) point out coating of cutting tools. Improved tribological properties promote reduced application of lubricating liquids, thus minimizing waste, providing a viable alternative to wet machining in the form of dry machining. From a technological perspective, high-speed machining requires improved heat and wear resistance due to local heat generation as well as heat-insulating properties of a tool coating (Santecchia et al., 2015). Typical properties of TiN coatings important for cutting tool applications can be listed as follows (*TiN Coatings*, 2021):

- Thickness 2–4 μm
- Friction coefficient ca. 0.5 (depends on application and test conditions)
- Hardness (HV 0.05) 2200 (depends on application and test conditions)
- Working temperature max. 600°C

Pogrebnjak et al. (2019) emphasize that multilayer nitride coatings exhibit excellent mechanical, optical, electronic, and magnetic properties, as well as improved durability and advantageous wear and corrosion resistance. They indicate specific combinations of functional characteristics of multilayer coatings related to the following:

1. Ability to provide final stresses of required sign and magnitude
2. Inhibition of heat flows due to frictional generation of heat during the cutting process
3. Limitation of interdiffusion processes between the coating and the substrate

After Pogrebnjak et al. (2019), some important examples of TiN-based multilayer coatings can be provided.

Titanium nitride-based multilayer coatings may consist of two layers, e.g., *TiN and TiC*, which contributes to a prolonged cutting tool lifetime compared to that of a single-layer TiN coating. Further, combinations like TiN, TiC, and/or TiCN with Al_2O_3 are found advantageous, since a layer of titanium nitride provides a high strength and the oxide layer improves chemical resistance. Each layer is commonly ca. 500 nm thick, providing an overall coating thickness of 5–10 μm.

Among superhard materials, *TiN/VN multilayer coating* can be named as one of the best examples, with hardness up to 56 GPa and nanoscale thickness of 5.2 nm. However, this coating exhibits susceptibility to oxidation, which poses some limitations to its applications. *ZrN/TiN multilayer coatings* of 1.5 μm thickness demonstrate 30% longer tool life in difficult milling conditions, compared with uncoated tools or tools coated with single ZrN layer.

Mechanical properties of multilayer coatings can be also improved by introduction of layers of new nanocrystalline or amorphous materials. For instance, *TiN/MoN coatings* exhibit high hardness of ca. 30 GPa, which is 25% higher than that of single-layer coatings, while their thickness is 25 nm (Pogrebnjak et al., 2019).

However, Łępicka et al. (2019) draw attention to the fact that even though TiN is considered an almost all-purpose coating, examples of adverse effects and wear intensification after TiN deposition have been reported, too. When a titanium nitride coating is placed on mild or low-alloy steels, TiAl intermetallics, or aluminum 1000 series, wear resistance becomes ambiguous and dependent on the applied load. The authors conclude that wear performance of metallic substrates coated with TiN depends on friction conditions and material properties, especially the elastic modulus mismatch between an outer coating and an underlying substrate. They also suggest that other factors should be taken into consideration, such as differences in chemical reactivity of the substrate and coating materials.

3 Smart Machining Processes

3.1 COMPUTER-INTEGRATED MANUFACTURING

During the development of industrial processes, mechanization facilitates mass production to meet the consumer demand for improved products. However, to achieve mass production, transfer lines and fixed automation are needed, leading to the development of programmable automation. Its main objective is to accelerate the production process throughout the factory and to increase the quality of products. As such, automation provides for quick response to changing customer requirements with high-quality products. With developments in commercially available information and communication technology tools and equipment, the application of computers to manufacturing begins, collectively named advanced manufacturing technologies (AMTs). In today's industrialized world, further enormous growth in manufacturing automation has brought a plethora of AMTs with a large variety of features, consisting of semi- to fully automated systems or equipment. However, individual automation in functional units produces many islands of automation in an enterprise that cannot facilitate communication between functional units. In addition, errors in data sharing and other mismatches with these automation units continually plague the manufacturing industry. Thus emerges the need for a holistic and systematic integration of individual automation solutions adopted by manufacturers (Nagalingam and Lin, 2008).

Nowadays, information and communication technology plays a significant role in manufacturing systems (Wang et al., 2021). The ongoing development of cyber systems and related intelligent and smart technologies has given rise to the concept of the so-called *Industry 4.0*, which includes big data processing, cloud computing, the Internet of Things (IoT), cyber-physical systems (CPSs), digital twin (DT), and next-generation artificial intelligence (AI). Various advanced manufacturing paradigms have been proposed, using these concepts to enhance manufacturing processes and systems with various degrees of "intelligence" or "smartness" (Wang et al., 2021).

Big data and analytics refer to the collection and comprehensive evaluation of data from many different sources, which consist of four dimensions: (1) volume of data, (2) variety of data, (3) velocity of generation of new data and analysis, and (4) value of data (Vaidiya et al., 2018).

A cloud-based IT platform serves as a technical backbone for linking and communication of manifold elements through connections of different devices to the same cloud to share information (Vaidiya et al., 2018).

The *Internet of Things* means a worldwide network of interconnected and uniform addressed objects that communicate via standard protocols. It is also referred to as the Internet of Everything (IoE) and consists of the Internet of Service (IoS),

DOI: 10.1201/9781003218654-3

the Internet of Manufacturing Services (IoMs), the Internet of People (IoP), an embedded system, and Integration of Information and Communication Technology (IICT). The three key features of the IoT are context, omnipresence, and optimization (Vaidiya et al., 2018).

Phuyal et al. (2020) emphasize that the use of the IoT has been adopted widely in the field of industrial automation in developed countries. The architecture of IoT is basically divided into four layers:

1. The first layer includes sensors and actuators which are integrated into hardware to collect information.
2. The networking layer serves transmission of collected data from sensors to control units and from control units to actuators.
3. To access services based on user needs and to interact with the control units, a service and interface layer is distinguished and defined. Information from a physical system is collected through the use of sensors and different machine learning tools. The collected information is then managed systematically, processed in local smart devices, and then sent to the application layer through the network layer for implementation in the respective fields.
4. The application layer decides to take necessary action in an IoT integrated system (Phuyal et al., 2020).

The term "cyber–physical system" (CPS) can be understood as a system where natural and human-made systems (physical space) are tightly integrated into computation, communication, and control systems (cyber space). Decentralization and autonomous behavior of the production process are the main features of the CPS, dependent on adoption and reconfiguration of product structure supply networks. The latter may be considered as collaborative cyber-physical systems used in manufacturing systems. Continuous intelligent interchanging of data is carried out by linking cyber-physical systems through cloud systems in real time. The basic requirement of real-time-oriented manufacturing operation and optimization of actual production systems is achieved by means of proper sensors in the CPS (Vaidiya et al., 2018).

A *digital twin* (DT) can be created as a result of the fusion of real and digital systems. When field data are used or transferred to a simulation model, a realistic digital image is created; validation and retrofitting of the predictive model are performed with realistic data. The simulation model learns from actual experiences and events in reality, which makes it no longer a simple theoretical calculation. This model reproduces the real behavior of a machine or individual components almost identically and is thus called "a digital twin" (Werner et al., 2019).

Next-generation artificial intelligence (Next-Gen AI) is a framework that incorporates explainability, interpretability, transfer learning, and ensemble learning with a wide variety of learning architectures. It uses the context of previously known information and facilitates the use of human knowledge and experience for experimental trial design and interpretation of results. Since the first-generation AI is being used in surveying and classifying omics data, it is designed to solve well-defined tasks of single-omics datasets and does not require the integration of data across multiple modalities. On the other hand, the next-generation AI can change the dynamics of how experiments are planned, thus enabling better data integration, analysis, and interpretation (Harfouche et al., 2019).

Galati and Bigliardi (2019) stress that the recent advancements in terms of information technology, electronics, and industrial robots have led to the Internet of Things (IoT), virtualization technologies, cloud computing, computer-integrated manufacturing (CIM) systems, and then cyber–physical systems. The latter enable production systems to be modular and easy to reconfigure, which is required to produce highly customized products in mass production. Thus, Industry 4.0 technologies can support companies in terms of several design principles like interoperability, virtualization, decentralization, real-time capability, service orientation, and modularity.

With the advent of the fourth industrial revolution (Industry 4.0), manufacturing systems experience a transformation whereby the IoT and other technologies play a major role in shifting manufacturing companies to computer-integrated manufacturing (Chen et al., 2020). Nagalingam and Lin (2008) name three main sources that induce diverse challenges to manufacturing enterprises:

- Technological changes
- Geopolitical and economic environment
- Sophisticated and demanding customers

Today's conditions of the open market, especially advancements and modern tools in information and communication technology (ICT), allow customers to reach and choose manufacturers of their preference across the world. For the global customers today, the primary determinants become rather non-price factors, such as quality, product design, innovation, and delivery services. Thus, an enterprise that relies on traditional manufacturing systems cannot satisfy the needs of globally distributed customers and match the capabilities of its competitors. Manufacturers should have the ability to be flexible, adaptable, proactive, and responsive to changes and should be able to produce a wide variety of high-quality and innovative products quickly and at a reduced cost. In addition, they should be able to meet new environmental requirements and complex social issues, set by dynamic geopolitical boundaries and conditions. This has led to the concept of virtual computer-integrated manufacturing (VCIM), which is a network of interconnected and globally distributed CIM systems across geographical boundaries, whereas CIM is an integration of localized manufacturing facilities. In 1998, the National Science Foundation of the United States attempted to create a vision of a competitive environment for manufacturing, publishing the *Visionary Manufacturing Challenges for 2020*, which identified the most important technical, political, and economic forces for manufacturing (Nagalingam and Lin, 2008):

- Sophisticated customers will demand products that are customized to meet their needs.
- Rapid responses to market forces are required to survive in the competitive climate, enhanced by communication and knowledge sharing.

- Creativity and innovation are required in all aspects of a manufacturing enterprise for it to be competitive.
- Developments in innovative process technologies will change both the scope and scale of manufacturing.
- Environmental issues will be predominant as the global ecosystem becomes strained by growing populations and the emergence of new high-technology economies.
- Information and knowledge will be shared by manufacturing enterprises and the marketplace for effective decision-making.
- Global distribution of highly competitive production resources will be a critical factor in the organization of manufacturing enterprises to be successful in this changing technical, political, and economic climate.

Essentially, CIM is a process of using computers and communication networks to transform technological units into a highly interconnected manufacturing system. Over the decades, interpretation of CIM gradually has changed from computerized work-cells, large-scale automation, CAD/CAM, interfacing and communication concepts to a contemporary mature definition where CIM is an integration effort that embraces a whole organization across all functional units. In order to achieve an effective integration, an in-depth understanding of all technologies and comprehensive knowledge of all activities across all functional units in an enterprise is required. Since computers are merely tools that facilitate processing and exchanging information, successful application of CIM requires, first, integration of functional units and software applications/systems. Today, with sophisticated manufacturing equipment and available advanced ICT tools, it is possible to reach the objectives of CIM with ease (Nagalingam and Lin, 2008).

While the concept of CIM is broadening its application to manufacturing and other industries, new terms have been coined by various academics, perhaps in an attempt at better understanding and implementation of CIM, and maybe to demonstrate efforts of the researchers/consultants to be innovative. Among these terms reflecting the dynamic nature of improvements in manufacturing applications are lean manufacturing (LM), just-in-time (JIT), cellular manufacturing, concurrent engineering, agile manufacturing, responsive manufacturing, holonic manufacturing, distributed manufacturing, and collaborative manufacturing. Although new terms like these are generated every few years, the concept of CIM is far broader and able to embrace all of them, providing the required features offered by these concepts or strategies. Therefore, CIM still stands as the innovative application for yesterday's issues, but also newer applications required today and for the future (Nagalingam and Lin, 2008).

The introduction of an integrated manufacturing system provides many potential benefits, including (Nagalingam and Lin, 2008):

1. Easier communication enabled between a functional unit of an enterprise and other relevant functional units
2. Accurate data transfer between a manufacturing plant and/or subcontracting facilities

3. Faster responses to required changes
4. Increased flexibility toward introduction of new products
5. Improved accuracy and quality in manufacturing processes
6. Improved quality of products
7. Effective control of dataflows among various units
8. Reduction of lead times
9. Streamlined manufacturing flow from order to delivery
10. A holistic approach to enterprise-wide issues.

The so-called *smart parts-based manufacturing systems* ensure that each part is correctly identified to perform reliable process control in such a flexible and customer-oriented manufacturing system. Uniquely identified individual parts can be processed according to their specific requirements based on individual customer preferences. Chen et al. (2020) describe how to use radio frequency identification (RFID) in computer-integrated manufacturing as a lot identification system with the following benefits:

- Low cost when compared with smart tag systems
- No need to detach tags while POD is cleaned
- Increased ease of carrier tracing
- Reduced probability of missing lots (when compared with smart tag systems)
- Ease of maintaining the system and minimizing tag failure rate
- Reduced complexity of CIM system
- Operation benefit with smart tag
- Ability to fulfill a "place and go" scenario
- Is an easier to operate split function when compared with barcodes

3.2 SMART MANUFACTURING

Owing to recent advances in information technologies (IT) and operation technologies (OT), the manufacturing industry has entered the era of Industry 4.0 and the traditional manufacturing technologies empowered by artificial intelligence and the Industrial Internet of Things (IIoT) have become "smart" at handling the ever-increasing complexity of tasks (Baroroh et al., 2020). Nowadays, it is imperative to implement smart manufacturing that offers high production flexibility and efficiency at acceptable costs; otherwise it is impossible to meet the rapidly growing consumer demand for mass customized products (Baroroh et al., 2020).

Over the past years, the topic of smart manufacturing has been the focus of researchers and manufacturing experts. Academic and industry researchers have developed two paradigms to describe the deep integration of manufacturing and advanced information technologies: smart manufacturing (SM) and intelligent manufacturing (IM) (Wang et al., 2021). The authors find that around 2000, IM was primarily associated with expert systems, fuzzy logic, neural networks, agent, flexible manufacturing systems, CIM, and computer-aided design (CAD), while the early phases of modern SM around 2010 are Industry 4.0 and automation. The most recent keywords for IM are Industrial Internet, smart factory, cloud computing, and CPSs, while the most recent keywords associated with SM are CPSs, smart factory, cloud computing, big data, and IoT. The authors conclude that expanding the application of Industry 4.0 concepts and practices is likely driving keyword usage in both SM and IM concepts.

In general, smart manufacturing describes the technology-driven ability of a system to solve both existing and future problems in a collaborative manufacturing infrastructure which responds to changing demands in real time. However, many industrial enterprises are still unsure what smart manufacturing entails and which potential benefits and challenges it holds (Zenisek et al., 2021). Maggi et al. (2021) state that the complexity of smart manufacturing systems makes it futile to provide any comprehensive definition of the concept of smart manufacturing itself. They only conclude that smart manufacturing systems are the modern implementation of the previous totally integrated automation (TIA) concept, while from the standpoint of cybersecurity research, smart manufacturing represents the frontier of industrial control systems (ICSs).

According to Wang et al. (2021), the application of intelligence to manufacturing has emerged as a compelling topic for researchers and industries around the world. While the terms "smart manufacturing" and "intelligent manufacturing" (IM) are similar, they are not identical. After a thorough bibliometric analysis of publication sources, annual publication numbers, keywords frequency, and top regions of research and development, the authors conclude that under various definitions, different concepts and research topics can be associated with SM or IM in different development phases. The development of digitalization, networking, and intelligentization in manufacturing is common for both paradigms. Since manufacturing enterprises are the main implementers of SM and IM, the authors suggest more attention should be paid to key technologies such as CPS, big data, cloud computing, IoT, and AI, and human/staff education must be undertaken based on their unique actual situation, no matter which of the two paradigms, SM or IM, is adopted (Wang et al., 2021). A better understanding of the potential of smart manufacturing also requires investigation of an additional subset of smart manufacturing technologies, including mixed reality, additive manufacturing, and predictive maintenance (Zenisek et al., 2021).

Maggi et al. (2021) draw attention to the fact that the concept of a "reference" smart manufacturing system does not really exist. They emphasize security issues are of paramount importance for both the cybersecurity and ICS communities. Despite the trend to integrate and interconnect, smart factory systems are still relatively closed, so there is little chance that conventional mass attacks will hit the "closed world" of the smart factory. However, the main drawback is that a single security flaw may allow an attacker to gain full access to a factory machine, posing a serious threat to the rest of the network (Maggi et al., 2021).

Decision-making is an important component of any intelligent system. According to Assad et al. (2021), it is a repetitive procedure that takes place in a manufacturing facility at all levels and is executed by humans or industrial controllers. In both cases, a "condition" that initiates the decision-making has to be reported and then a corrective action is started. The following points are widely recognized in the relevant state of the art (Assad et al., 2021):

- Condition monitoring is of vital importance for the maintenance process.
- The recent trend is to employ the IoT for data collection so that further processing and decision-making are possible.
- Condition monitoring is also useful for production control.
- Much work has been done in the field of machining, but less for assembly lines.
- Considerations of life cycle assessment in smart manufacturing are not given enough attention.
- New opportunities exist under Industry 4.0.

Phuyal et al. (2020) summarize the main characteristics and challenges of smart manufacturing systems:

- Security issues in smart manufacturing
- System integration (machine-to-machine communication and interconnectivity of a system require a better communication system)
- Interoperability (ability of different systems to understand and access each other's functions independently) at four levels of Industry 4.0, namely operational, systematical, technical, and semantic
- Safety in human–robot collaboration – foremost consideration to be given to occupational health and safety of personnel working on the site, to avoid any hazardous environment, and to maintain necessary occupational health and safety
- Multilingualism – ability to interpret any instructions given in a human language into a machine language to instruct a machine on the desired operation (AI implementation)
- Return on investment in new technology

To investigate implementation paths of the smart manufacturing information system (SMIS), Zhang and Ming (2021) analyze 42 representative enterprises from key fields, including offshore engineering equipment manufacturing, aerospace equipment manufacturing, household electrical appliances manufacturing, 3C manufacturing, automobile manufacturing, shipbuilding, medical equipment manufacturing, agricultural equipment bureau manufacturing, manipulator equipment manufacturing, and rail transit equipment manufacturing industries. The smart characteristics dimension and system layer dimension are assessed in four implementation directions: (1) resource-based automated factories, (2) network-based interconnection factory, (3) platform-based data sharing factory, and (4) information system integration factory. The authors conclude the implementation path for SMIS is usually divided into six big steps, each of them divided into five sub-steps. Further divisions are derived from the five-level manufacturing system dimension and five-level product life cycle dimension, improved to customize manufacturing based on customer needs. The implementation steps can be described as follows (Zhang and Ming, 2021):

Step 1 is the establishment of a resource-based automated factory, including five sub-steps: equipment for core technology and key short board, digital simulation and research and development (R&D), automatic production line, digital workshop or factory, and intelligent management within an enterprise.

Step 2 is the establishment of a network-based interconnection factory, which includes five sub-steps of vizualization: equipment and energy, process and quality, monitoring and visualization of production and logistics, monitoring and visualization of workshops, and large screen for visual display of enterprises.

Step 3 is the establishment of platform-based data sharing factory, which includes five sub-steps: data acquisition and management platform for equipment, management platform of simulation data, platform of production line data, management platform of production data, monitoring and visualization of workshops, and data management platform for product life cycle.

Step 4 is the establishment of an information system integration factory, which includes five sub-steps: information system integration for equipment hierarchy, process hierarchy, production line hierarchy, workshop or factory hierarchy, and enterprise hierarchy.

Step 5 is the establishment of a new model factory for the product life cycle, which includes five sub-steps: innovating new models of development and design, planning and scheduling, flexible production, intelligent logistics, and sale and service.

Step 6 is the establishment of the new model factory for personalized customization, which includes five sub-steps: innovating new models of dynamic demand, personalized customization, open collaborative design, flexible manufacturing, and experiential service.

Although AI enables manufacturing systems to operate with a high degree of autonomy and intelligence, there are some tasks impossible to accomplish without human intervention. Thus, according to Baroroh et al. (2020), the purpose of introducing AI or automation technologies is not to completely replace human involvement, but rather to facilitate manual operations. In tasks involving huge data or fuzzy conditions, AI may not perform better, because humans can utilize their cognitive capabilities or implicit knowledge to respond quickly. Thus, to let humans and machines work with each other, in a complementary fashion, seems to be a more feasible approach. The ideas of "humans in the loop" or human–cyber–physical system (HCPS) reflect this objective. In this context, integrating augmented reality (AR) with intelligent functions can be considered a good strategy. AR can serve as an interface that strengthens interactions between a human operator and manufacturing environment allowing the former to assess the ambient intelligence through the AR interface and ensuring his proper response to manufacturing tasks in real time.

Based on their literature survey, the authors find the most frequent implementation of AR-assisted tools takes place in the manufacturing operation of assembly/disassembly. A majority of deployed AR applications involve multiple solution functions and intelligence sources to deal with current complex manufacturing problems. Visual clue/perception is a major sensory channel to communicate with users of AR. The objective is to increase the human operator's situational awareness in the manufacturing environment. The authors conclude that AR is an interfacing technology able to exchange information with humans in real time and its technical merits become evident when AR is applied to tasks in which humans and computational intelligence can complement each other (Baroroh et al., 2020).

Emerging digital technologies of Industry 4.0 provide applicability of the Internet of Things, virtual reality (VR), and augmented reality in remanufacturing. A study by Kerin and Pham (2019) suggests there is still a need to explore the connection of cyber-physical systems to the IoT to support smart remanufacturing, while aligning with evolving information and communication infrastructures and circular economy business models.

Smart devices like the IoT and CPS have now emerged as a universal paradigm that can drastically transform any industries equipped with sensing, identification, remote control, and automated control capabilities. The concepts and programs like Industry 4.0, Society 5.0, Made in China 2025, and Industrial Internet are all based on the internet and interconnected devices and basically have as their theme process control through less human intervention and smart decisions, with a huge impact on the global market (Phuyal et al., 2020).

3.3 "SMART" TOOLS AND MATERIALS

Cheng et al. (2017) express the opinion that smart tooling has tremendous potential as a generation precision machining technology of particular importance in the Industry 4.0 context. Zhao, Liu et al. (2021) point out that a new field of smart cutting tools has emerged from the integration of sensors into traditional cutting tools to reduce chatter, to measure cutting force and cutting temperature, and to monitor tool wear and damage. Möhring et al. (2020) divide monitoring methods into two areas:

1. Direct monitoring via camera systems or microscopes
2. Indirect monitoring through force measurement, acoustic emission, and vibrations

In their review, Hopkins and Hosseini (2019) indicated three main directions in the development of smart metal cutting tools, namely:

1. Self-regulation of regenerative vibration (chatter) through monitoring of vibrations, predicting stability lobes of a milling tool, semi-active damping control, or process simulations.
2. Monitoring of work conditions in order to determine the point at which a tool becomes no more useful, to prevent excessive increase of the cutting forces, or to predict tool wear. Cutting forces, temperature, and changes in spindle acceleration are among the measured parameters.

3. Signal interpretation. To make a tool "smart," i.e., able to react adequately to work conditions, transmission and integration of signals are required. In this respect, the possibility of different sensors' communication with each other during machining seems to be the most promising, as well as self-learning probabilistic neural network. In ultrahigh-precision micromilling, a smart machining platform is able to significantly reduce vibrations.

Möhring et al. (2020) developed a smart milling tool with indexable inserts that comprises a cyber-physical system. It is based on wear and force measurements with determined VB values of wear land at the flank face and SB cutting edge misalignment at the rake face. The data obtained from the measurements are filtered, converted, and stored, and then compared with the reference process and previously collected measurement data. This way, the user receives all the necessary data to determine the service life and wear conditions of a tool.

Smart cutting tools have autonomous sensing and self-learning capabilities and are able to operate in-process sensoring and actuation (Cheng et al., 2017). With these capabilities, they ensure improved quality of a machined part and its surface roughness, reduced fabrication costs, and higher manufacturing productivity. The following characteristics of smart cutting tools should be listed (Cheng et al., 2017):

- Plug-and-play
- Autonomous operation
- Self-condition monitoring
- Self-positioning adjustment
- Self-learning
- Compatibility with highly automated CNC environments

The authors (Cheng et al., 2017) describe in detail four types of smart toolings they developed, namely:

1. Cutting force measurement-based smart cutting tool, which employs a piezoelectric film or a surface acoustic wave sensor as the sensoring element.
2. Cutting temperature-oriented smart cutting tool, focused on a controlled internal cooling system, reducing the cutting temperature around the cutting edge in order to extend tool life and produce a better surface finish on the workpiece.
3. Fast tool servo employed to position a cutting tool operating in a dynamic cutting and actuation scenario with a high precision accuracy and wide bandwidth, for precision machining of complex geometrical features in particular.
4. Smart fixtures and smart collets to measure cutting force, cutting temperature, tool positioning and actuation in process, separately or combined. Smart collets and smart fixtures are essential as they enable smart machining by machining system devices.

The term "smart materials" was coined at the beginning of the 21st century. This sort of material is a logical stage in the development of materials utilization by humans (Pupan and Kononenko, 2008), as shown in Figure 3.1.

As underlined by Bahl et al. (2020), smart materials cannot be defined in a single specific way. They are normally defined as advanced or intelligent materials that can respond smartly to environmental changes. The authors categorize smart materials on the basis of their properties, distinguishing passive smart materials with the ability to transfer a type of energy and active materials that are also divided into two categories. Smart materials that cannot change their properties when exposed to external stimuli belong to the first type, while the second type includes materials that can turn one form of energy (thermal, electrical, chemical, mechanical, and optical) into another form. A schematic of this categorization highlighting the amazing properties of the active smart materials is shown in Figure 3.2.

Filimon (2019) stresses it is crucial to understand the origin and determinants of the behavior of smart materials that result from a molecular adjustment to changes caused by the external field. In modeling this behavior, she distinguishes two approaches, namely micromechanical and phenomenological, which represent the material behavior regardless of its origin and provide a basis for dividing smart materials into the following two groups:

1. Materials that undergo changes in one or more of their properties (chemical, mechanical, electrical, magnetic, or thermal) in direct response to a change of external stimuli. These changes are direct and reversible with no need for an additional control system to cause them to occur. Among the materials in this category are thermochromic, magnetorheological, thermotropic, and shape memory alloys (SMAs).
2. Materials that transform energy from one form to output energy in another form and again, directly and reversibly. For instance, an electrorestrictive material transforms electrical into mechanical energy changing its physical shape. Among the materials in this category are piezoelectrics, thermoelectrics, photovoltaics, pyroelectrics, and photoluminescents.

Mukherjee et al. (2021) list eight groups of smart materials, as follows:

1. *Piezoelectric materials* change their electrical properties when a force is applied to them, showing both converse and direct effects.
2. *Electrostrictive materials* are similar to piezoelectric materials with respect to their mode of action, but the difference lies in a proportional change to

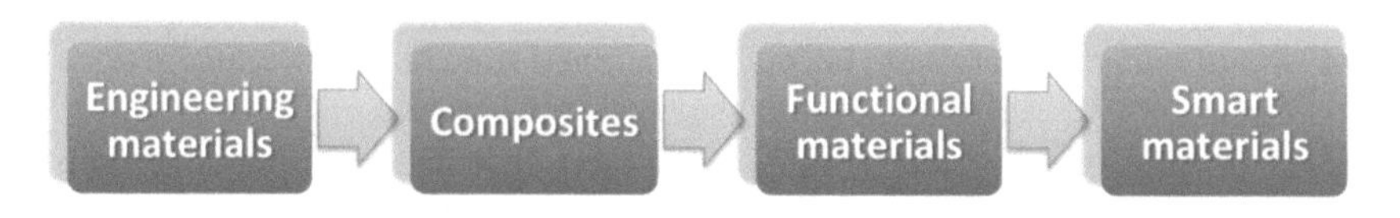

FIGURE 3.1 Development of materials.

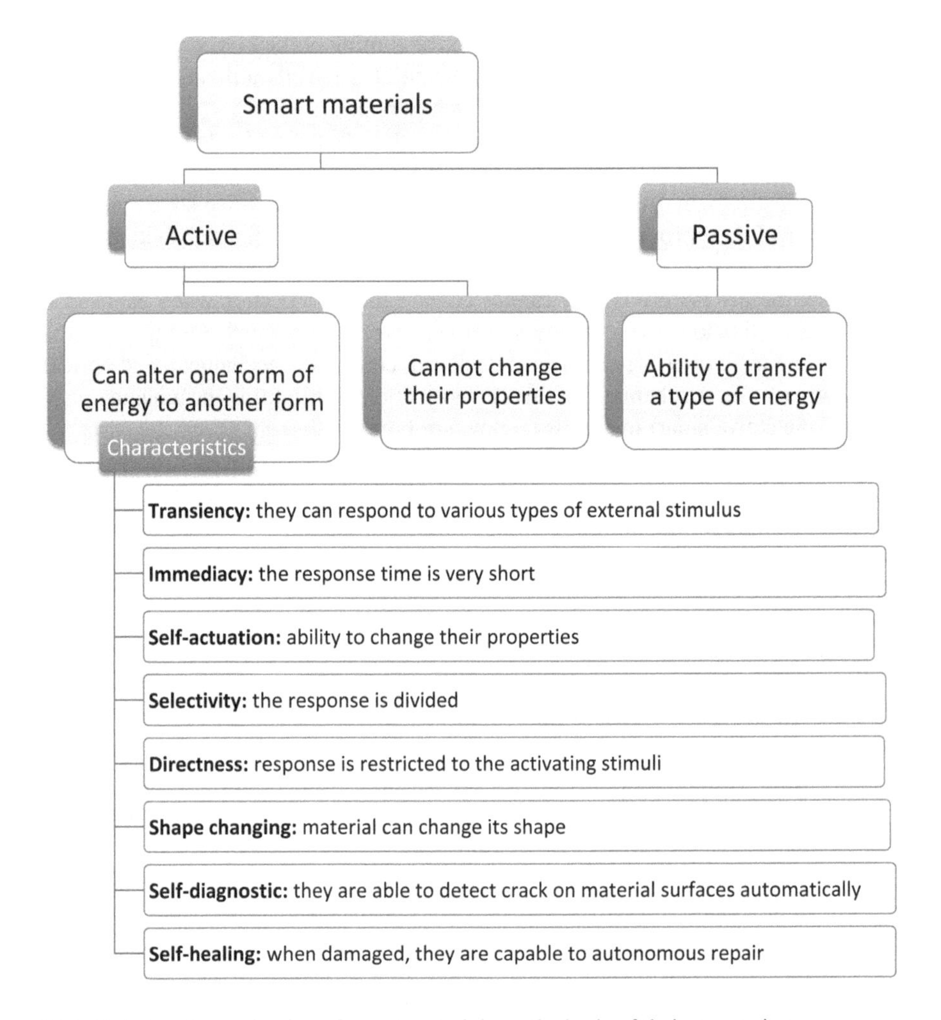

FIGURE 3.2 Categorization of smart materials on the basis of their properties.

the order of 2 degrees. When an electric charge is applied, all the particles change their positions and rearrange in a certain direction, improving the electrical properties of the material.

3. *Magnetostrictive materials* show a change in the strain state when a magnetic field is applied and vice versa. Similar to ferromagnetic materials, all particles of a magnetostrictive material get aligned in accordance with the magnetic field applied.
4. *Rheological materials* are normally in a liquid state but when a magnetic or electrical field is applied, they change their physical state.
5. *Thermo-responsive materials* are polymers that drastically change their physical properties when exposed to any temperature variation. They are also called stimuli-responsive materials.

6. *Electrochromic materials* are an imperative part of today's liquid crystal display units. Such a material changes its optical properties when an electric current is passed through it.
7. *Fullerenes* are allotropes of carbon made up of hexagons and pentagons connected by single and double bonds forming a caged sphere with carbon atoms at its nodes. Due to its shape, fullerene is highly stable and versatile in nature.
8. *Biomimetic materials* are inspired by nature. Their geometric shapes are very simple yet effective in terms of purposes like strength, camouflage, water proofing, and mobility. Examples of such materials can be honeycomb structures, coconut tree leaves, bird's flight, beehive, spider web, water repellency of fish skin, etc. Their applications range from self-sensing to self-repairing in buildings.

Hu (2016) characterizes a number of active materials for textile coating, namely smart and polymeric hydrogels, memory polymers, phase-change materials, color-change materials, and functional nanomaterials.

Smart or polymeric hydrogels belong to a group of hydrogels that display various changes under specific external conditions such as temperature, pH, light, salt, and stress. These materials respond with swelling/collapsing and hydrophilic/hydrophobic changes in shape. Among the hydrogels most commonly used in active coating are temperature-active hydrogels, where transition temperature is adjusted by additives modifying monomer structure or copolymerization.

Memory polymers are smart materials that can sense thermal, mechanical, electric, and magnetic stimuli and respond by changing their shape, position, stiffness, and other static and dynamic characteristics. They have low costs and exhibit good processing ability and controllable responses, making them even more suitable for industrial production than SMAs. Becoming the most widely applicable smart materials, this group of memory polymers have developed rapidly in both academic and industrial areas.

Adaptive polymeric particles include nanoparticles and microcapsules of different chemical and physical parameters, such as morphology, shape, size, light reflection/diffraction, and solvent ability. Smart materials combined with particle materials provide beneficial and unique properties to textile coating owing to their tiny forms and responsive characteristics different from normal particles. The surface properties of nanoparticles are more essential than those of microcapsules and applications of active coating of these particles include self-cleaning textiles, phase-change microencapsulation textiles, and hydrophilic/hydrophobic textiles (Hu, 2016).

Phase-change materials (PCMs) possess the ability to change their state within a certain temperature range, absorbing energy during the heating process as phase changes take place and transferring it to the environment in the phase change range during a reverse cooling process. The insulation effect reached by the PCM takes place only during the phase change in a certain temperature range and terminates when the phase change in all of the PCMs is complete. Hence, this type of thermal insulation is temporary and can be referred to as dynamic thermal insulation (Mondal, 2008).

Materials that change color are termed "chromogenic" or "chromotropic," sometimes also called chameleonic because they reversibly change color as a response to changes in environmental conditions (Ferrara and Bengisu, 2014). In certain applications, a permanent color change is preferred, which is also possible with some chromogenic materials. The technical principle by which these materials function can be explained by an alteration in the equilibrium of electrons caused by a stimulus, with a consequent modification of reflectance, absorption, emission, or transmission. This process, named chromism, involves a change in the microstructure or electronic state of substances, mostly in conjugated polymers (Ferrara and Bengisu, 2014).

Among the most interesting *functional nanomaterials*, Fe–Pd ferromagnetic alloys are worthy of attention due to their excellent functional properties such as high uniaxial magnetic anisotropy, high Kerr rotation, magnetic shape memory effect, high corrosion resistance, and biocompatibility. Prida et al. (2012) underline that in addition to thin films, Fe–Pd nanowires and antidots are also promising smart materials suitable for magnetic shape memory nanoactuators, magnetocaloric or high-density data storage micro-devices.

Guo et al. (2020) describe a group of smart materials with the ability to sense four common biomolecules, namely glucose, nucleic acids, proteins, and enzymes. The authors point out that smart materials, such as hydrogels, or various nanomaterials like gold nanoparticles, quantum dots, carbon nanotubes, and graphene have been widely used for applications in biosensing. Photonic crystals, i.e., periodic micro- or nanostructures that can control the reflection of light, provide an excellent platform for biosensing with visible readout to report target analytes. Molecularly imprinting polymers comprise a class of materials with a special recognition site to bind with imprinted molecules used for detection of various biomolecules such as proteins and enzymes, based on responsive changes in refractive index and volume. Especially sensitive materials, such as electrochemical, optical, thermal, piezoelectric, and impedimetric sensors, as well as interferometric biosensors have been widely employed as diagnostic platforms.

3.4 SUPERPLASTICITY AND ITS APPLICATION TO METAL FORMING

According to Humphreys et al. (2017), *superplastic materials* are polycrystalline solids that have the ability to exhibit unusually large ductilities and often elongations of several hundred percent when deformed in uniaxial tension, and exceptionally, even up to several thousand percent, in contrast to a normal ductile metal that will fail by necking after an elongation of less than 50%. This is illustrated in Figure 3.3 with two strain-stress diagrams. The authors name a wide variety of metals, alloys, and even ceramics that can be made superplastic. Production of superplastic microstructures in any crystalline material exploits the principles of recrystallization and grain growth, as well as stability of such a microstructure during subsequent high-temperature deformation (Humphreys et al., 2017).

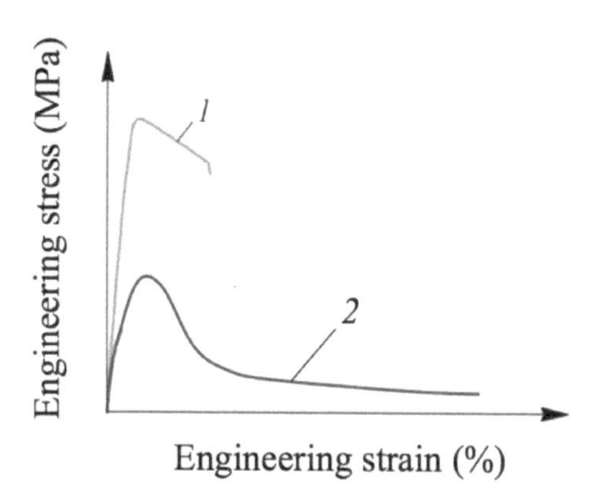

FIGURE 3.3 Simplified stress-strain diagrams for typical engineering (1) and superplastic materials (2).

In fact, there are three main requirements for manifestation of superplasticity in microcrystalline materials, namely (Dudina et al., 2016):

1. Fine and equiaxed grains whose size is reasonably stable during deformation
2. Temperature that is more than about half the melting point of the matrix in absolute degrees
3. A strain rate that is not too high or too low, i.e., between 10^{-2} and 10^{-6} s^{-1}

Moreover, Dudina et al. (2016) point out that with the refinement of grain size to nanoscale, superplasticity can be extended to lower temperatures and significantly higher strain rates.

Zhang (2010) states the existence of second phases is the best way to stabilize the grain size, because a small amount of second phase particles in alloys can inhibit the grain growth by pinning grain boundaries. However, the second phase particles are unable to hinder the grain growth completely. Thus, the optimum microstructure for superplasticity appears to be a "microduplex" structure, where ultrafine grains of two or more phases are arranged alternately. Eutectic and eutectoid alloys often have this microduplex characteristic, hence their microstructure is very stable during deformation (Zhang, 2010).

Superplasticity can be classified as follows (Harwani et al., 2021):

1. Internal stress superplasticity (ISS) that involves developing internal stresses in a material by thermal cycling, which later helps achieve large strains during deformation
2. Fine structural superplasticity (FSS), also known as micro-grain superplasticity, which relies on the fine grain microstructure that remains thermally stable during large tensile deformations

Ren et al. (2019) underline that the mechanisms responsible for superplastic deformation have long remained a controversial subject. The proposed theories include grain boundary sliding, grain rotation, grain rearrangement, dislocation activity, diffusional creep, and dynamic recrystallization, on the assumption that an actual superplastic deformation should involve a combined, not a single, mechanism. Depending on the sigmoidal relationship between stress and strain rate, the superplastic flow behavior is divided into three mechanism zones (Langdon, 1982):

1. Diffusion creep (region I)
2. Grain boundary sliding (region II)
3. Dislocation creep (region III)

At present, a consensus has been reached on the dominant mechanism of grain boundary sliding in region II at intermediate strain rates, where it accounts for 50% to 80% of the total deformation during the superplastic deformation process. This ratio varies with material, composition, grain size, and temperature (Ren et al., 2019).

It has been found that at high temperatures, a superplastic alloy is characterized by a low flow stress below 10 MPa and a high resistance to nonuniform thinning, which allows for near-net-shape forming of sheet material using techniques similar to those for forming of thermoplastics (Humphreys et al., 2017). *Superplastic forming* (SPF) is an attractive option for forming complex shapes from the sheet at low stresses with substantially reduced tooling costs compared to conventional cold-pressing operations. However, the slow strain rates often necessary for superplastic forming and the higher material costs have so far restricted the commercial exploitation of SPF to a relatively small number of specialized applications (Humphreys et al., 2017). Pupan and Kononenko (2008) highlight spectacular results of investigations on deformation increase, such as elongation close to 8,000% in case of aluminum bronze, more than 800% for zirconia ceramics, and 1,400% for metal matrix composites. The authors also draw attention to increasing deformation rate and provide examples of superplasticity at rates typical for traditional metal forming processes including explosive forming. This sort of high-strain-rate superplasticity has been observed in aluminum-based and magnesium-based alloys (10^{-2} to 10^{-1} s^{-1}). For ceramic materials consisting of tetragonal zirconium oxide, magnesium aluminate spinel, and α-alumina phases, superplasticity is achieved at strain rates of up to 1 s^{-1} (Kim et al., 2001).

Process modeling has been widely used in the SPF industry in order to perform proper optimization of complex forming processes, possible only with a precise characterization of superplasticity. At present, there are two recognized forms of constitutive models that describe superplastic behavior, namely mechanism-based and phenomenological constitutive equations. The mechanism models mentioned in most references are based on the grain boundary sliding phenomenon with one or more coordination mechanisms (Ren et al., 2019).

Pupan and Kononenko (2008) point out that superplastic forming is effective in the following conditions:

- When non-ductile alloys are processed, impossible to be subject to plastic deformation in normal conditions
- When a formed part consists of thin elements such as reinforcing ribs or thin sheets

- When exceptional accuracy with minimum waste is required, e.g., due to the high price of a starting material
- When forging large components that require increased power in the normal forging process, or components of elevated quality and safety requirements

Superplasticity forming is now a widely accepted method for producing vanes of gas turbine engines out of nickel-based superalloys and for forming airframes for military applications (Dudina et al., 2016). There is widespread interest worldwide in applying SPF to forming automotive parts, but since a much-desired factor in the heavily competitive sector of automobile production is the reduction of forming time, intense research is aimed at refining grain size that leads to faster forming rates. Perhaps SPF is most used for sheet materials (Pupan and Kononenko, 2008). A schematic of superplastic sheet forming is shown in Figure 3.4. Due to low flow stress of superplastic deformation, gas pressure between 0.1 and 2.1 MPa appears to be a more effective means of forming than stamping dies that require expensive mechanical equipment.

The following methods of sheet SPF may be distinguished (Pupan and Kononenko, 2008):

- Negative and reverse pneumatic forming
- Vacuum forming
- Pneumatic forming with a movable or immobile punch

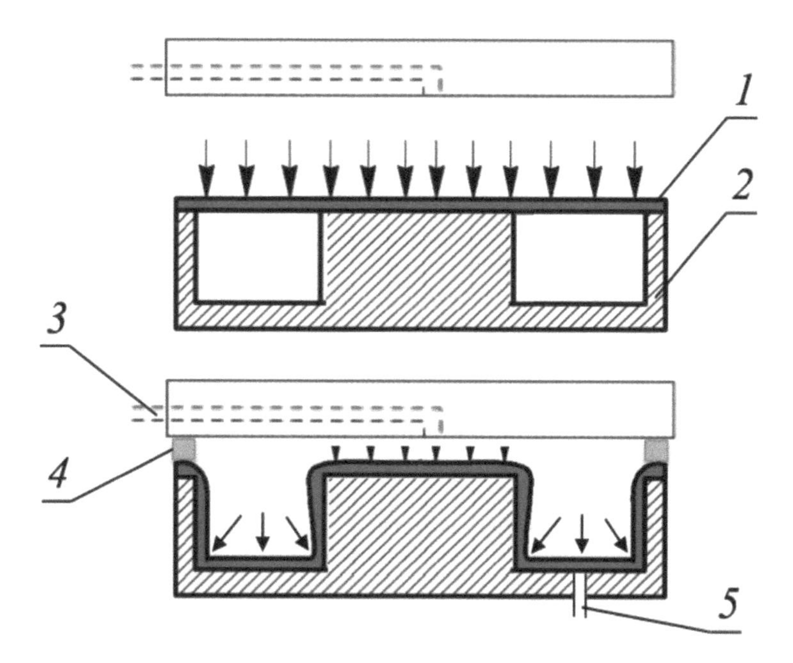

FIGURE 3.4 Schematic of superplastic sheet forming: 1 – Metal sheet, 2 – Die, 3 – Gas pressure, 4 – Seal bead, 5 – Vent.

The basic steps involved in superplastic forming of metal sheets include the stages when a sheet is placed within a die cavity, then heated while high gas pressure is evenly applied, causing plastic deformation of a metal at very large strains into a complex-shaped, single-piece component (Mouritz, 2012), as illustrated in Figure 3.4. As a rule, SPF involves slowly forming a sheet of material in a single-sided tool at a temperature of about 500°C during a typical forming cycle, which for SPF is approximately 30 minutes or even longer (Powell et al., 2012).

Apart from sheet forming, SPF can be realized in other traditional metal forming processes, such as rolling, drawing, and precise volumetric stamping. In particular, SPF is advantageous in the following cases (Pupan and Kononenko, 2008):

1. Empty, *complex shape and deep stretching* can be obtained from a metal sheet in one operation with superplastic deformation, while traditionally several operations are required. Such components can be made of aluminum and titanium alloys, matrix metal composites, and even ceramic materials.
2. *Large and complex machine components*, e.g., panels with wafer structure.
3. *Tool forming*, such as molds, punches, dies for stamping, etc. Some high-speed steels used for tools possess fine crystalline structures and require little additional efforts to put them into the superplasticity state with optimal temperature and strain rate. On the other hand, carbon tool steels must undergo initial thermal or thermomechanical treatment before making them superplastic. The authors provide examples of reamers, cutters, and countersinks made of R6M5 steel (Russian nomenclature, corresponding to HS6-5-2 in Europe and T11302-M2 in the United States) in superplastic conditions of 810–820°C, strain rate 10^{-1} s^{-1}, and specific deformation pressure 180–250 MPa.
4. *Ceramics and intermetallic compounds* can be made superplastic due to the specific preparation of a starting material and sintering conditions that produce a fine-structured crystalline bulk material. As a result, ceramics can be processed by forging or extrusion with elongation of 120–150%, while in normal conditions they cannot stand elongation above 3%. Apart from Al_2O_3, ZrO_2, and their composites, a sort of bioactive ceramic $Ca_{10}(PO_4)_6(OH)_2$ is found to exhibit superplasticity. This material, known as calcium hydroxyapatite (HA), is one of the most important implantable materials by analogy to mineral components of natural bones, used as a substitute material for human hard tissues. Aluminides of titanium and nickel are the most interesting intermetallic compounds.

Li et al. (2020) report unique superplasticity within a supercooled liquid region in *metallic glasses* (MGs), making them ideal materials for precise and net-shaping of various geometries by thermoplastic forming and breaking through limitations of poor processability of bulk metallic glasses (BMGs) at ambient temperature. BMGs are used for miniature fabrication in structural and functional applications, so that superplasticity and small solidification shrinkage in the supercooled liquid region make it possible to obtain a variety of micro-/nano-surface structures or three-dimensional (3D) parts of BMGs with high fidelity. For instance, the width of grooves formed by SPF is 18.45 ± 0.05 μm with a very small average error of ca. 0.27% related to the silicon mold, deviation of formed from designed angles is less than 0.1°, and the depth of BMG channels in different areas is highly repeatable (Li et al., 2020).

Among the most promising technologies that are based on superplasticity, superplastic forming/diffusion bonding (SPF/DB) can be named (Li et al., 2015). SPF/DB uses material superplasticity and diffusion to manufacture hollow or honeycomb structures of high complexity in a single step. Major advantages of this technology include high design freedom, short lead time, little spring-back, reduced number of parts, and increased structure integrity. It is also cost-saving, since 10–50% weight savings and 25–40% cost reduction can be achieved by using an SPF/DB structure. Such structures made of Al, Ti, and superalloys are mainly used as aerospace components such as ducts, aircraft wing access panels, rudders, nozzles, engine casings, and blades. A single-layer SPF structure has been used as a rocket fuel tank with wall thickness as low as 1.8 mm and variation within 0.15 mm. SPF/DB has significantly reduced welding work in this sort of fuel tanks and increased structural integrity. In optimization of SPF/DB structure, numerical methods are used to simulate stress, strain, and thickness distribution, while processing parameters are analyzed to predict and avoid risks during processing. Future developments include simulation of SPF/DB forming process, prediction of life-cycle-long properties of produced components, and optimization of structural parameters. These will have a large impact on application of SPF/DB structures in the aerospace industry (Li et al., 2015).

3.5 SEVERE PLASTIC DEFORMATION IN NANOSTRUCTURAL MATERIALS PROCESSING

Nanocrystalline (NC) materials with a grain size in the range of 1–100 nm have emerged as a new class of materials with unusual structures (Mohamed and Li, 2001). Because of such characteristics, these materials exhibit unique microstructures in which the volume of grain boundary is significant and affects various physical, electrical, chemical, magnetic, and deformation properties in the ultrafine grain size range. One of the most important features of NC materials is an interesting possibility to address the low strain rates of the superplastic region, usually 10^{-5} to 10^{-2} s^{-1}, too slow to be economically suitable for a variety of industrial applications. Mohamed and Li (2001) emphasize that as the grain size decreases from micrometer to nanometer, the superplastic region can be transposed to high strain rates or observed at lower temperatures, exhibiting a high strain rate and/or low-temperature superplasticity. Wang, Jiang et al. (2020) report properties of bulk nanocrystalline Mg fabricated by cryomilling and spark plasma sintering with an average grain size of 74 nm. They note superplastic strain of ~120% during compression at room temperature with a strain rate of up to 10^{-2} s^{-1}, as well as an increased strain rate sensitivity.

The recent development of third-generation materials is characterized by their ultrafine-grained or NC structure produced via severe plastic deformation (SPD), represented by high-pressure torsion and equal channel angular pressing (ECAP), or by bottom-up synthesis like electrodeposition in various alloys and even pure metals (Masuda and Sato, 2020). These techniques enable low-temperature and/or high-strain-rate superplasticity with grain growth restricted by a reduced deformation temperature and time.

In the *high-pressure torsion* (HPT) process, a specimen is held between a plunger and support and is strained in torsion under an applied pressure p in the order of 1–10 GPa (Borodachenkova et al., 2017). As shown in Figure 3.5, the lower holder (3) rotates causing deformation of the specimen (2) by the contact surface friction forces under a quasi-hydrostatic pressure. Torsion under such pressure results in the reduction of the grain size (Pupan and Kononenko, 2008).

In practice, there are two main types of HPT processing distinguished according to the shape of the anvils, namely unconstrained and constrained. In *unconstrained HPT*, samples are placed between two anvils and their material is free to flow outward when the pressure and torsion are applied. As a result, much thinner elements can be obtained than in *constrained HPT*, where specimens are placed into a cavity of the lower anvil or both anvils that prevent a material from flowing outward. However, the constrained HPT is a more common method (Borodachenkova et al., 2017). The main disadvantage of this method is limited dimensions and forms of processed specimens that must have diameters between 10 and 20 mm and thickness 0.2–0.5 mm (Pupan and Kononenko, 2008).

Among modifications of the HPT methods, Segal (2018) points out the following :

- High-pressure torsion of long samples, where long billets are processed using an incremental or continuous material transfer through a torsion zone
- High-pressure tube twisting, in which a sample is compressed by two punches between a die and mandrel and is then additionally twisted by a rotating die
- High-pressure sliding, where sheet material is compressed between anvils and sheared by the movement of the lower anvil in forward and backward directions, instead of the rotationary movement shown in Figure 3.5.

The *equal channel angular pressing* (ECAP) is the most popular and widely studied SPD technique schematically shown in Figure 3.6. As described by Kapoor (2017), a workpiece with a square or round cross section is pressed into a die with two intersecting channels of an equal cross-section area. The workpiece is pressed with a plunger and all its material exits through the second channel, experiencing shear strain in the intersection area of the two channels. Since the cross section is symmetric, either square or circular, the sample can be rotated before reentry into the die. Passing the sample through an ECAP die produces a simple shear deformation, each pass through a die with a 90° angle will result in an effective strain of 1, and n such passes will result in an effective strain of n (Kapoor, 2017).

The fine nanocrystalline structure can be obtained after two or three passages through the die, but usually ca. eight passages are required to obtain uniform grains of dimensions 100 nm and smaller (Pupan and Kononenko, 2008). ECAP can be

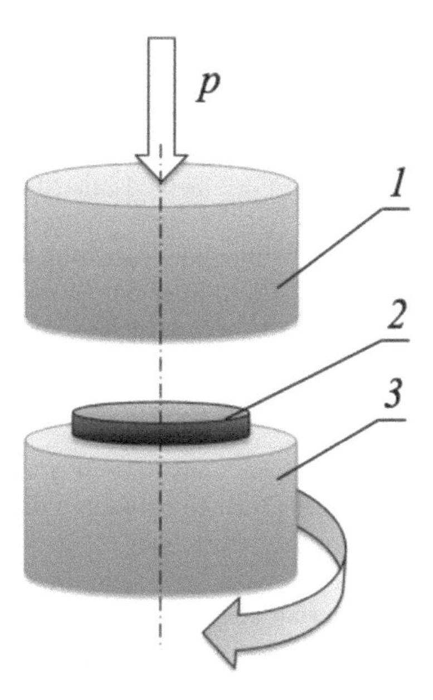

FIGURE 3.5 A schematic of the HPT principle: 1 – Upper anvil, 2 – Specimen, 3 – Lower anvil.

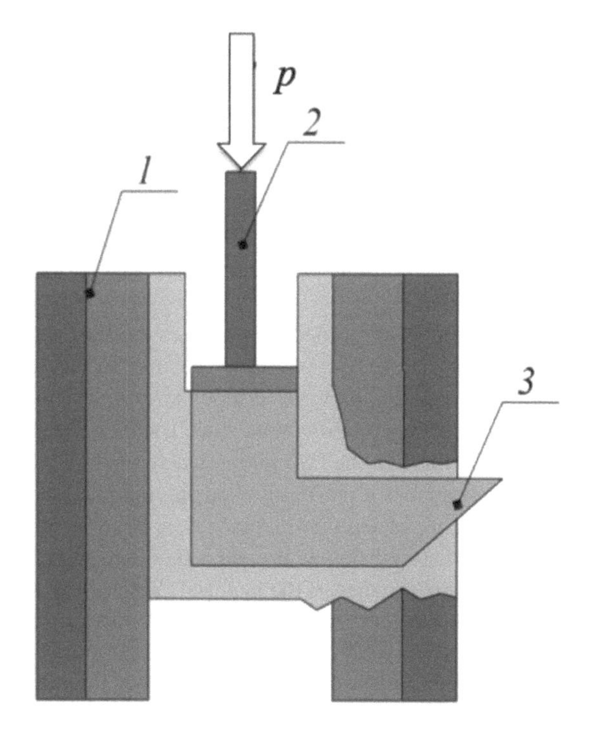

FIGURE 3.6 A schematic of the ECAP principle: 1 – Die, 2 – Plunger, 3 – Workpiece.

applied both to pure metals including hard-to-deform ones, such as tungsten or titanium, and to alloys. In the case of difficult-to-deform metals and alloys, the ECAP process is performed at elevated temperatures. It should be noted that the conditions such as time and temperature of the subsequent passages may be varied.

There are also several modified ECAP methods, listed below (Segal, 2018):

- Continuous ECAE-conform method for processing of long materials
- Semi-continuous incremental-ECAP method, where material shearing and feeding are separated for small successive steps to reduce press capacity, making processing of long billets possible
- Multi-pass ECAE in rotary dies or at special presses, without billet ejection after each pass
- Tubular-channel angular pressing consisting of a three-step passage of a pipe sample through channels formed between a die and mandrel, which can only be applied to relatively short tubes

Superplastic forming of NC materials at low homologous temperatures or high strain rates has been among the first examples of applying NC materials forming as a near-net-shape production process. It is a favorable shape-forming method, since complex shapes can be obtained without involving multistep processing routes and thus in a cost-effective manner (Wilde, 2014).

Friction stir processing (FSP), an adaptation of the solid-state joining process friction stir welding (FSW) described in Section 1.13.2, also belongs among widely recognized SPD techniques. Friction stir processing with three consecutive passes causes a dramatic decrease in the grain size from 23 ± 2 μm of the base material of Mg–Y–Nd alloy to 2.1 ± 1 μm, about 90% of the grains are converted to an ultrafine grain microstructure below 1 μm after the second FSP pass in 6-mm thick magnesium silver earth (QE22) specimens, while 80% of the grains are changed to ultrafine structure after the first pass, and for AZ31 Mg alloy two-pass FSP (1,000 rpm with a travel speed of 37 mm/min for both the passes) results in nano-grains of average size 85 nm (Harwani et al., 2021).

Unlike FSW, where samples are joined, FSPutilizes the same process principles to modify the local microstructure of monolithic specimens (Węglowski, 2018). As can be seen in Figure 3.7, a rotating tool travels along a workpiece to achieve specific and desired properties by surface modification, since frictional heat and mechanical stirring lead to grain refinement in the FSP region where the tool has passed.

Like in FSW, the tool induces a plastic flow during the friction stir process, but depending on the selection of process parameters, i.e., applied force, traveling speed, and rotational speed, the material flow can yield a modified microstructure that is beneficial to the required performance of the material (Węglowski, 2018). The width of the processed zone after a single pass of the tool is slightly wider than the diameter of a pin.

The grain size reduction is achieved by increasing tool rotation speed and a simultaneous decrease in the tool traverse speed *v*. Higher tool rotation accounts for a faster breakup of grains combined with greater mechanical intermixing of particles,

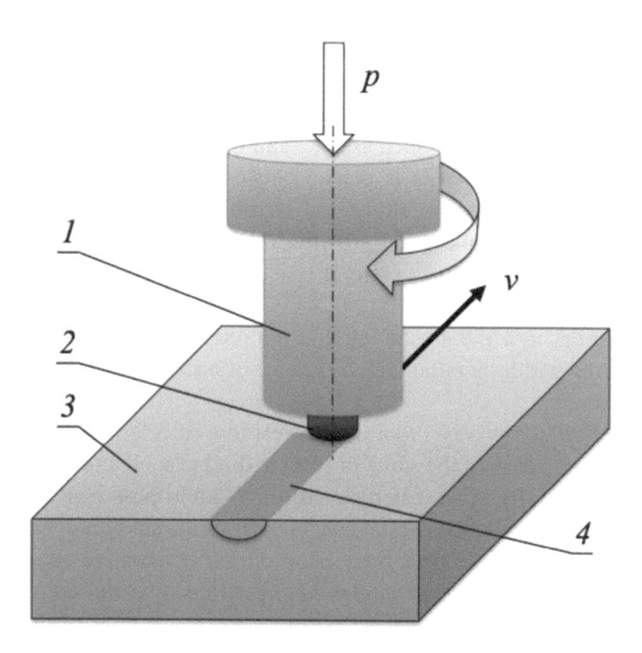

FIGURE 3.7 A schematic of the FSP principle: 1 – Shoulder, 2 – Pin, 3 – Workpiece, 4 – FSP region.

which leads to finer and more uniformly distributed grains and may generate second-phase particles during the FSP process (Harwani et al., 2021). The authors also point out that the tool configuration has a profound influence on heat generation, material flow, and temperature distribution during FSP, which is precisely reflected in the material grain structure.

The thermomechanical nature of FSP involves a combination of high temperature and strain, which modifies the initial microstructure in several ways. There are three distinct microstructural zones (Bauri and Yadav, 2018), namely:

1. Stir zone (SZ), the region which experiences maximum temperature, strain, and maximum grain refinement. SZ is also called a nugget zone or a dynamically recrystallized zone and is characterized by equiaxed fine grains. The SZ's volume depends on the tool dimensions. Due to nonuniform deformation during processing, the advancing side of SZ has a relatively sharper interface between the SZ and an adjacent thermomechanically affected zone (TMAZ) than the retreating side. The grain sizes exhibit variations from top to bottom and also across the zone from advancing to the retreating side.
2. Thermomechanically affected zone (TMAZ) appears around the volume of the SZ and experiences lower temperature compared to SZ. The strain rate in TMAZ is too low to cause dynamic recrystallization. It contains deformed and elongated grains with high dislocation density inside. The

size of this zone varies from several micrometers to millimeters dependent on the material and transition from the SZ to the TMAZ is usually not sharp. Soft metals, e.g., aluminum and magnesium, have wider zones compared to harder materials such as steel and titanium.
3. Heat-affected zone (HAZ) is placed between TMAZ and base material. This zone experiences no strain and is subjected to only thermal cycles which can lead to some grain growth. Mechanical properties in HAZ are poor, causing failures during tensile tests.

The zones exist on both the advancing and the retreating side; however, they may not have the same width on either side. The size of each zone depends on a number of process parameters.

Single-pass FSP comprises a tool that travels in a selected area only once, but if the tool is traversed again in the same or reverse direction on a previously processed area, wholly or partially, then it is referred to as "multi-pass FSP." Harwani et al. (2021) underline that multi-pass FSP can furnish grain size reduction and texture weakening more than single-pass processing due to cumulative strain rates in the successive passes and also enhances the homogeneous distribution of grains in the material. In fact, FSP performs selective superplastic forming in desired regions. Gangil et al. (2021) describe FSP as a technique for strengthening of high specific strength alloys, including age-hardened aluminum alloys, for several industrial applications through surface composite fabrication, where friction stir processing generates reinforcement particles.

Węglowski (2018) points out that FSP, like other FSW techniques, can be carried out using both specialized and typical milling machines. The former are equipped with dedicated measurement systems, whereas the latter need a universal measurement head. However, milling machines generally fail to provide ideal conditions for FSP due to an important issue of independent measurement systems development.

The ability of FSP and FSW to provide improved microstructural and mechanical properties by the grain refining process has given rise to an alternative method of solid-state (non-melting) process named *friction stir-based additive manufacturing* (FSAM) (Srivastava et al., 2021). The process has many benefits, such as solid-state nature, wrought microstructural properties, flexibility in material feed, and capability to process almost any type of metals and alloys. Due to low temperatures, FSAM can produce components with negligible porosity, inclusions, particle segregation, internal cavities, hot cracking dilution, etc., which make it an alternative for most beam-based MAM methods. Srivastava et al. (2021) point out that FSAM is a rapidly growing technique, executed by joining various materials layer upon layer to develop a 3D object from digital data, working on the principle of friction stir welding. The main difference is that joining of a layer by another layer which is associated with reheating and re-sintering of a material and flexibility to adjust the reheating and re-sintering time allow the process to control the microstructural features, so that desired mechanical properties can be obtained.

As emphasized by Węglowski (2018), other areas of possible FSP applications include increasing materials plasticity, modification of welded joints, production of composite materials, repair of casting elements, etc. It should be noted that FSP is a green technology producing no fumes and dusts, so perhaps in the coming years, it will become increasingly important and competitive relative to traditional surface engineering technologies.

3.6 PRODUCT LIFE CYCLE MODELS AND POTENTIAL CULTURAL OR SOCIAL IMPLICATIONS OF REPAIR

Decisions made by an engineer at the early stages of product development can result in unintended consequences that have a propagation effect during product life-phases of manufacturing, use, and disposal (Borg and Giannini, 2003). Under the pressure of market competition and environmental care, major attention is paid during product development from the very design stages to all life-cycle issues. The concept of a "product life cycle" has been widely discussed since the late 1950s with numerous models proposed (Polli and Cook, 1969). The life cycle model usually comprises a series of subsequent stages guided by management decisions which qualify a system to progress to its next life-cycle stage (Pena et al., 2021). In the context of increasing demand for manufacturing sustainability with respect to energy, process, material, and environment friendliness, the life cycle models are critical in the assessment of the performance of a product from the design phase to its end of life (EoL) (Daniyan et al., 2021). It is noteworthy that among the articles reviewed by these authors, 60% applied life cycle assessment (LCA) methodology to reduce energy consumption and enhance environmental sustainability, while 40% employed other assessment tools. The most important approaches found in literature can be listed as follows (Daniyan et al., 2021):

1. LCA principles for *low-carbon manufacturing* are focused on energy savings during design and manufacturing, use, transportation, and remanufacturing stages of the product life cycle.
2. Life cycle inventory (LCI) and *impact assessment*, where environmental impacts, energy and materials consumption, as well as other variables are measured and recorded.
3. *Reconfiguration* of manufacturing systems as an effective way of redesigning components of a system to ensure efficiency, modularity, scalability, and flexibility. This approach encourages optimized costs of fabrication of quality products while ensuring efficient use of resources with minimal impact on the natural environment.
4. *Internet of Things* and web-based applications which enable technical communications related to design, manufacturing process parameters, and technical support.
5. *Computer-aided process planning* and cloud platforms allow for the identification of factors which influence the degradation of machining resources, compensating them and minimizing waste generation, as well as improving energy efficiency and production time.
6. *Life-cycle-oriented services* aimed at improving resource efficiency of a machine tool.

7. *Life cycle costing and profit approach* assumes that the optimum cost of procurement will ensure the effective performance of a machine tool without sacrificing quality and performance. Estimation and prediction of life cycle cost and management of machine tools are helpful in decision-making.
8. *Preventive structuring approach* is based on product design. During the early design stage, an engineer can weigh the possibility of effects caused by a combination of chosen technologies on processing accuracy.
9. *Tool and equipment management approach* that involves machining process monitoring and tool wear detection.
10. *Technical product service* system.
11. *Real-time monitoring approach* which is achieved by incorporation of sensors for monitoring of machine availability and quality of machining processes.
12. *Green cutting technology* can be obtained by proper selection of cutting fluids and their right application, thus enhancing machine tool life and sustainability.
13. *Remanufacturing approach* can improve eco-friendliness of manufacturing and reduce the consumption of non-renewable resources.
14. *Modeling approach* which combines the life cycle assessment and the design of experiments (DoE).
15. *Multi-criteria decision approach* that considers various economic, environmental, and social indicators but also takes into account the recovery process of material after its end of life.

Further improvement of a product's sustainability can be achieved through the life cycle gap analysis (Dieterle and Viere, 2021). It is a specific technique to interpret LCA results from a circular economy (CE) perspective. Technically, the product's life cycle gap (LCG) describes the difference between the environmental impacts of a product's initial manufacturing and its environmental value after recycling. This approach measures and visualizes the distance between the vision of CE aimed at the closed loop of a product's material and energy flows and the status quo, without ignoring the shifting of burdens from one life-cycle stage to another. The authors suggest integration of social assessments into LCGA is another promising future research strategy.

In this respect, it is very interesting to pay attention to a relatively rare study on potential cultural, social, or personal implications of the repair and circular economy conducted by McLaren et al. (2020). The authors emphasize that in circular economy literature, repair and restoration are typically understood as good tools to extend product life spans and reduce waste, while sociological repair literature reveals its contrasting understanding in terms of reconciliation and reconfiguration. The instrumental and technocratic view of repair diverges from that found in sociological, ethnographic, and political studies examining repair and maintenance of material objects and infrastructures, which suggest a broad diversity of intentions, motivations, and values. It is of particular interest to the authors whether application of "repair ethics" to CE practices might transform or "reconfigure" rather than sustain or "restore" the industrial society.

In the literature reviewed by McLaren et al. (2020), there is a huge predominance of implicit instrumental understandings of repair, present in more than four-fifths of all the analyzed papers mentioning remanufacturing or repair. This understanding is focused on extending product life, increasing resource efficiency, or reducing and preventing waste, rarely bringing cultural or emotional considerations to the center. However, as many publications indicate, a circular economy tends to introduce important social transformations, including changes in labor relations, involving an end to consumerism or an end to material growth, which would be needed to deliver sustainability. Thus, the emerging sociology of repair is focused on social and political values and systems contesting the instrumental approach to repair issues. McLaren et al. (2020) consider three key dimensions of this sort of contestation:

- Is repair a *sustaining or transforming* process?
- Is the temporal focus of repair backward- or forward-looking?
- Is the activity of repair personal or political?

They underline that these are not simple dualisms, but posing the questions in this way they intend to identify deeper implications of repair beyond the prevailing instrumental narrative in CE. After all, decisions about repair can be seen as choices of how to engage with these or similar binary questions, since repair is not some abstract objective process, but a way of acting on the world which cocreates cultural values, social and economic relations, and material outcomes. Since repair is necessarily a relational process, it affects social relations and identities directly or indirectly as much as it has material consequences. The authors conclude their analysis as follows:

> Research into repair in circular economies therefore needs to become more inter- and trans-disciplinary, linking political, cultural, economic and technological investigation. In particular, researchers could do much more to understand non-instrumental motivations for repair, especially considering individual actors as more than consumers; to examine the relational, social, moral and cultural implications of business models such as product service systems; to assess who wins and who loses from different configurations of repair; and to explore the political economy of business approaches to repair and circularity, as well as the potential for policy mechanisms that could reframe repair, for instance as part of degrowth, or within novel forms of capitalism or even non-capitalist social relations.
>
> **(McLaren et al., 2020)**

Thus, any engineer making his decisions on design, material, and processing technologies of future products should keep in mind their huge impact on the natural environment and society both in local and global terms.

3.7 CONCLUDING REMARKS

Today's engineers much more than ever are forced to deal with multiple systems, subsystems, and supra-systems in a large variety of domains. It is enough just to mention the number of available engineering materials, which is close to 100,000 and far beyond human ability to keep them all in mind usefully. Alteration of materials' properties in processing, compositions of various structures in micro- and nanoscale, and surface engineering methods provide endless combinations impossible to describe in full.

Therefore, the presented processes and techniques are far from a complete review but give an insight into recent trends. These are pushed forward by increasing demand for new materials with enhanced properties and additional functions, but constrained by limited natural resources and environmental impact of processing and numerous life-cycle phases of a product, including its reuse, remanufacturing, or recycling. The concept of circular economy has become reality with its regulations that have an effect on engineering activities. Thus, we decided to describe some of the materials and methods in the context of the product life cycle, mostly focusing on remanufacturing issues.

Another important context of the presented technologies is the CIM and especially the Industry 4.0 concept, which includes big data processing, cloud computing, industrial IoT, additive manufacturing systems, cyber–physical systems, digital twins, and Next-Gen AI. Smart materials, smart tools, smart processes, and eventually smart factories not only improve the effectiveness of engineers' activity but also have a huge potential for taking over control and regulations in manufacturing processes.

We believe that our book is helpful in understanding the main relations and links between various systems of materials, their properties, processing parameters, and product life-cycle issues. This understanding is crucial to engineers' decision-making and finding good support for the decisions made at various stages of the product life cycle. It should be kept in mind that engineers are responsible not only for the safe work of a particular component or engineering system but for its long-term impact on ecology, energy, and resource savings, as well as social life.

References

Abramczyk, H. 2005. *Introduction to Laser Spectroscopy.* Amsterdam: Elsevier. https://doi.org/10.1016/B978-044451662-6/50005-8

Adalarasan, R., M. Santhanakumar, and M. Rajmohan. 2015. Application of Grey Taguchi-based response surface methodology (GT-RSM) for optimizing the plasma arc cutting parameters of 304L stainless steel. *The International Journal of Advanced Manufacturing Technology* 78:1161–1170. https://doi.org/10.1007/s00170-014-6744-0

Alexenko V.L., N.V. Brailo, and N.V. Serdiuk. 2016. Protection of the steel workpieces from oxiadtion at high temperatures during heat treatment and press forming. *Scientific Bullettin of the Kherson State Maritime Academy* 1(14):137–145 (in Russian).

Allain, J.P., and M. Echeverry-Rendón. 2018, Surface treatment of metallic biomaterials in contact with blood to enhance hemocompatibility. In: Ch.A. Siedlecki (Ed.), *Hemocompatibility of Biomaterials for Clinical Applications*, pp. 279–326. Cambridge, MA: Woodhead Publishing. https://doi.org/10.1016/B978-0-08-100497-5.00008-2

Allen D.M., P. Shore, R.W. Evans, C. Fanara, W. O'Brien, S. Marson, and W. O'Neill. 2009. Ion beam, focused ion beam, and plasma discharge machining. *CIRP Annals: Manufacturing Technology* 58(2):647–662. https://doi.org/10.1016/j.cirp.2009.09.007

Annicelli, R.A. 2020. Chemical blowing agents. In: J.S. Dick (Ed.), *Rubber Technology* (Third Edition), pp. 685–698. Cincinnati, OH: Hanser. https://doi.org/10.3139/9781569906163.021

Arnold, Th., and F. Pietag. 2015. Ion beam figuring machine for ultra-precision silicon spheres correction. *Precision Engineering* 41:119–125. https://doi.org/10.1016/j.precisioneng.2015.03.00

Assad, F., S. Konstantinov, H. Nureldin, M. Waseem, E. Rushforth, B. Ahmad, and R. Harrison. 2021. Maintenance and digital health control in smart manufacturing based on condition monitoring. *Procedia CIRP* 97:142–147. https://doi.org/10.1016/j.procir.2020.05.216

Attarzadeh, N., M. Molaei, K. Babaei, and A. Fattah-alhosseini. 2021. New promising ceramic coatings for corrosion and wear protection of steels: A review. *Surfaces and Interfaces* 23:100997. https://doi.org/10.1016/j.surfin.2021.100997

Averin, E. 2017. Universal method for the prediction of abrasive waterjet performance in Mining. *Engineering* 3(6):888–891. https://doi.org/10.1016/j.eng.2017.12.004

Baek, D.K., T.J. Ko, and S.H. Yang. 2013. Enhancement of surface quality in ultrasonic machining of glass using a sacrificing coating. *Journal of Materials Processing Technology* 213(4):553–559. https://doi.org/10.1016/j.jmatprotec.2012.11.005

Bahl, Sh., H. Nagar, In. Singh, and Sh. Sehgal. 2020. Smart materials types, properties and applications: A review. *Materials Today: Proceedings* 28(Part 3):1302–1306. https://doi.org/10.1016/j.matpr.2020.04.505

Balaji, P.S., and V. Yadava. 2013. Three dimensional thermal finite element simulation of electro-discharge diamond surface grinding. *Simulation Modelling Practice and Theory* 35:97–117. https://doi.org/10.1016/j.simpat.2013.03.007

Barmashenko, B.D., V. Rybalkin, A. Katz, and S. Rosenwaks. 2003. An extremely efficient supersonic chemical oxygen-iodine laser. In: 2003 Conference on Lasers and Electro-Optics Europe (CLEO/Europe 2003), Munich, Germany, pp. 128. https://doi.org/10.1109/CLEOE.2003.1312189.

Baroroh, D.K., Ch.-H. Chu, and L. Wang. 2020. Systematic literature review on augmented reality in smart manufacturing: Collaboration between human and computational intelligence. *Journal of Manufacturing Systems*, in press. https://doi.org/10.1016/j.jmsy.2020.10.017

Basiev, T.T., P.G. Zverev, and S.B. Mirov. 2003. Colour centre lasers. In: C. Webb, and J. Jones (Eds.), *Handbook of Laser Technology and Applications*. Vol. 2, pp. 499–522. Boca Raton, FL: CRC Press. https://doi.org/10.1201/NOE0750309608

Batchelor A.W., L.N. Lam, and M. Chandrasekaran. 2011. *Materials Degradation and Its Control by Surface Engineering*. London: Imperial College Press.

Bauri, R., and D. Yadav. 2018. *Metal Matrix Composites by Friction Stir Processing*. Oxford, UK: Elsevier.

Bellotti, M., J.R. De Eguilior Caballero, J. Qian, and D. Reynaerts. 2021. Effects of partial tool engagement in micro-EDM milling and adaptive tool wear compensation strategy for efficient milling of inclined surfaces. *Journal of Materials Processing Technology* 288:116852. https://doi.org/10.1016/j.jmatprotec.2020.116852

Benelmekki, M., and A. Erbe. 2019. Nanostructured thin films–background, preparation and relation to the technological revolution of the 21st century. In: M. Benelmekki, and A. Erbe (Eds.), *Nanostructured Thin Films*. Vol. 14, pp. 1–34. Amsterdam: Elsevier. https://doi.org/10.1016/B978-0-08-102572-7.00001-5

Berghaus, J.O., B. Marple, and C. Moreau. 2006. Suspension plasma spraying of nanostructured WC-12Co coatings. *Journal of Thermal Spray Technology* 15:676–681. https://doi.org/10.1361/105996306X147072

Bergs, T., M. Schüler, M. Dadgar, T. Herrig, and A. Klink. 2020. Investigation of waterjet phases on material removal characteristics. *Procedia CIRP* 95:12–17. https://doi.org/10.1016/j.procir.2020.02.319

Bernard, B., A. Quet, L. Bianchi, A. Joulia, A. Malié, V. Schick, and B. Rémy. 2017. Thermal insulation properties of YSZ coatings: Suspension plasma spraying (SPS) versus electron beam physical vapor deposition (EB-PVD) and atmospheric plasma spraying (APS). *Surface and Coatings Technology* 318:122–128. https://doi.org/10.1016/j.surfcoat.2016.06.010

Bhattacharyya, B. 2015. *Micro and Nano Technologies, Electrochemical Micromachining for Nanofabrication, MEMS and Nanotechnology*. Oxford, UK: William Andrew Publishing. https://doi.org/10.1016/B978-0-323-32737-4.00002-5

Bhattacharyya, B., and B. Doloi. 2020. *Modern Machining Technology*. London: Academic Press. https://doi.org/10.1016/B978-0-12-812894-7.00005-0

Biesuz, M., Th. Saunders, D. Ke, M.J. Reece, Ch. Hu, and S. Grasso. 2021. A review of electromagnetic processing of materials (EPM): Heating, sintering, joining and forming. *Journal of Materials Science & Technology* 69:239–272. https://doi.org/10.1016/j.jmst.2020.06.049

Blume, S., Ch. Herrmann, and S. Thiede. 2018. Increasing resource efficiency of manufacturing systems using a knowledge-based system. *Procedia CIRP* 69:236–241. https://doi.org/10.1016/j.procir.2017.11.126

Böllinghaus Th., G. Byrne, B.I. Cherpakov, E. Chlebus, C.E. Cross, B. Denkena, U. Dilthey, T. Hatsuzawa, K. Herfurth, and H. Herold. 2009. Manufacturing engineering. In: K.-H. Grote, and E.K. Antonsson (Eds.), *Springer Handbook of Mechanical Engineering*, pp. 523–786. New York: Springer.

Borg. J.C., and F. Giannini. 2003. Exploiting integrated 'product' & 'life-phase' features. In: R. Soenen, G.J. Olling (Eds.), *Feature Based Product Life-Cycle Modelling*, pp. 1–18. Boston, MA: Kluwer Academic Publishers.

Borodachenkova, M., W. Wen, and A.M. de Bastos Pereira. 2017. High-pressure torsion: Experiments and modeling. In: M. Cabibbo (Ed.), *Severe Plastic Deformation Techniques*. London: InTechOpen. https://doi.org/10.5772/intechopen.69173

Boujelbene, M., B. El-Aoud, E. Bayraktar, I. Elbadawi, I. Chaudhry, A. Khaliq, A. Ayyaz, and Z. Elleuch. 2021. Effect of cutting conditions on surface roughness of machined parts in CO2 laser cutting of pure titanium. *Materials Today: Proceedings* 44 (Part 1):2080–2086. https://doi.org/10.1016/j.matpr.2020.12.179

Bouzakis K.D., and N. Michailidis. 2018. Physical vapor deposition (PVD). In: S. Chatti, and T. Tolio (Eds.), *CIRP Encyclopedia of Production Engineering*. Berlin – Heidelberg: Springer. https://doi.org/10.1007/978-3-642-35950-7_6489-5

Brecher, C., F. du Bois-Reymond, J. Nittinger, T. Breitbach, D. Do-Khac, M. Fey, and S. Schmidt. 2016. Qualifying multi-technology machine tools for complex machining processes. *CIRP Journal of Manufacturing Science and Technology* 13:1–14. https://doi.org/10.1016/j.cirpj.2015.11.001

Cai, S., B. Yao, Q. Zheng, Zh. Cai, W. Feng, B. Chen, and Z. He. 2020a. Dynamic grinding force model for carbide insert peripheral grinding based on grain element method. *Journal of Manufacturing Processes* 58:1200–1210. https://doi.org/10.1016/j.jmapro.2020.09.029

Cai, W., J. Wang, P. Jiang, L. Cao, G. Mi, and Q. Zhou. 2020b. Application of sensing techniques and artificial intelligence-based methods to laser welding real-time monitoring: A critical review of recent literature. *Journal of Manufacturing Systems* 57:1–18. https://doi.org/10.1016/j.jmsy.2020.07.021

Campbell, J. 2015. *Complete Casting Handbook* (Second Edition). Oxford, UK: Butterworth-Heinemann. https://doi.org/10.1016/B978-0-444-63509-9.00018-2

Canas, E., M.J. Orts, A.R. Boccaccini, and E. Sanchez. 2019. Microstructural and in vitro characterization of 45S5 bioactive glass coatings deposited by solution precursor plasma spraying (SPPS). *Surface and Coating Technology* 371:151–160. https://doi.org/10.1016/j.surfcoat.2018.12.057

Cao, J., X. Chen, X. Zhang, Y. Gao, X. Zhang, and S. Kumar. 2020. Overview of remanufacturing industry in China: Government policies, enterprise, and public awareness. *Journal of Cleaner Production* 242:118450. https://doi.org/10.1016/j.jclepro.2019.118450

Cao, X., D. Yu, M. Xiao, J. Miao, Y. Xiang, and J. Yao. 2016. Design and characteristics of a laminar plasma torch for materials processing. *Plasma Chemistry and Plasma Processing* 36:693–710. https://doi.org/10.1007/s11090-015-9661-6

Caristan, Ch.L. 2004. *Laser Cutting Guide for Manufacturing*. Dearborn, MI: SME & AFFT.

Cattini, A., D. Bellucci, A. Sola, L. Pawłowski, and V. Cannillo. 2013. Suspension plasma spraying of optimised functionally graded coatings of bioactive glass/hydroxyapatite. *Surface and Coatings Technology* 236:118–126. https://doi.org/10.1016/j.surfcoat.2013.09.037

Celanese Website. 2021. https://www.celanese.com/engineered-materials/Technical-Information/Technical-Services/Processing/MuCell.aspx (accessed on February, 4, 2021).

Celik-Aktas, A., J.F. Stubbins, and J.M. Zuo. 2007. Electron beam machining of nanometer-sized tips from multiwalled boron nitride nanotubes. *Journal of Applied Physics* 102:024310. https://doi.org/10.1063/1.2757007

Chandrasekar, M., D. Shivalingappa, and Dr. Channankaiah. 2016. Recent developments in cladding process: A review. *International Journal for Innovative Research in Science & Technology* 2(10):310–315.

Chang, Sh., A. Liu, Ch.Y.A. Ong, L. Zhang, X. Huang, Y.H. Tan, L. Zhao, L. Li, and J. Ding. 2019. Highly effective smoothening of 3D-printed metal structures via overpotential electrochemical polishing. *Materials Research Letters* 7(7):282–289. https://doi.org/10.1080/21663831.2019.1601645

Changdar, A., and Sh.Sh. Chakraborty. 2021. Laser processing of metal foam - A review. *Journal of Manufacturing Processes* 61:208–225. https://doi.org/10.1016/j.jmapro.2020.10.012

Chavoshi, S.Z., and X. Luo. 2015. Hybrid micro-machining processes: A review. *Precision Engineering* 41:1–23. https://doi.org/10.1016/j.precisioneng.2015.03.001

Chen, Y.Q., B. Zhou, M. Zhang, and Ch.M. Chen. 2020. Using IoT technology for computer-integrated manufacturing systems in the semiconductor industry. *Applied Soft Computing* 89:106065. https://doi.org/10.1016/j.asoc.2020.106065

Cheng, J., Ch. Liu, Sh. Shang, D. Liu, W. Perrie, G. Dearden, and K. Watkins. 2013. A review of ultrafast laser materials micromachining. *Optics & Laser Technology* 46:88–102. https://doi.org/10.1016/j.optlastec.2012.06.037

Cheng, K., Z.C. Niu, R.C. Wang, R. Rakowski, and R. Bateman. 2017. Smart cutting tools and smart machining: Development approaches, and their implementation and application perspectives. *Chinese Journal of Mechanical Engineering* 30:1162–1176. https://doi.org/10.1007/s10033-017-0183-4

Choi, J.P., B.H. Jeon, and B.H. Kim. 2007. Chemical-assisted ultrasonic machining of glass. *Journal of Materials Processing Technology* 191(1–3):153–156. https://doi.org/10.1016/j.jmatprotec.2007.03.017

Chow, W.W., and S.W. Koch. 1999. *Semiconductor-Laser Fundamentals: Physics of the Gain Materials*. Berlin: Springer.

Chowdhury, S., N. Yadaiah, D.A. Kumar, M. Murlidhar, C.P. Paul, Ch. Prakash, G. Królczyk, and A. Pramanik. 2021. Influence of tack operation on metallographic and angular distortion in electron beam welding of Ti-6l-4V alloy. *Measurement* 175:109160. https://doi.org/10.1016/j.measurement.2021.109160

Chu, P.K., J.Y. Chen, L.P. Wang, and N. Huang. 2002. Plasma-surface modification of biomaterials. *Materials Science and Engineering: R: Reports* 36(5–6):143–206. https://doi.org/10.1016/S0927-796X(02)00004-9

Cong W., and Z. Pei. 2013. Process of ultrasonic machining. In: A. Nee (Ed.), *Handbook of Manufacturing Engineering and Technology*. London: Springer. https://doi.org/10.1007/978-1-4471-4976-7_76-1

Cooke, Sh., K. Ahmadi, S. Willerth, and R. Herring. 2020. Metal additive manufacturing: Technology, metallurgy and modelling. *Journal of Manufacturing Processes* 57:978–1003. https://doi.org/10.1016/j.jmapro.2020.07.025

Copertaro, E., F. Perotti, P. Castellini, P. Chiariotti, M. Martarelli, and M. Annoni. 2020. Focusing tube operational vibration as a means for monitoring the abrasive waterjet cutting capability. *Journal of Manufacturing Processes* 59:1–10. https://doi.org/10.1016/j.jmapro.2020.09.040

Courrol, L.C., R.E. Samad, L. Gomes, I.M. Ranieri, S.L. Baldochi, A.Z. Freitas, S.P. Morato, and N.D. Vieira Jr. 2006. Color center production by femtosecond-pulse laser irradiation in fluoride crystals. *Laser Physics* 16:331–335. https://doi.org/10.1134/S1054660X06020216

Cui, H., F. Lu, K. Peng, X. Tang, and Sh. Yao. 2010. Research on electron beam welding of in situ TiB2p/ZL101 composite. *Journal of Shanghai Jiaotong University (Science)* 15:479–483. https://doi.org/10.1007/s12204-010-1036-9

Curry, N., K. VanEvery, T. Snyder, and N. Markocsan. 2014. Thermal conductivity analysis and lifetime testing of suspension plasma-sprayed thermal barrier coatings. *Coatings* 4(3):630–650. https://doi.org/10.3390/coatings4030630

Daniyan, I., Kh. Mpofu, B. Ramatsetse, and M. Gupta. 2021. Review of life cycle models for enhancing machine tools sustainability: Lessons, trends and future directions. *Heliyon* 7(4):e06790. https://doi.org/10.1016/j.heliyon.2021.e06790

Darthout, E., G. Laduye, and F. Gitzhofer. 2016. Processing parameter effects and thermal properties of $Y_2Si_2O_7$ nanostructured environmental barrier coatings synthesized by solution precursor induction plasma spraying. *Journal of Thermal Spray Technology* 25:1264–1279. https://doi.org/10.1007/s11666-016-0450-4

Deng, T., J. Li, and Zh. Zheng. 2018. Micro-beam plasma polishing of ground alloy steel surfaces. *Procedia Manufacturing* 15:1678–1686. https://doi.org/10.1016/j.promfg.2018.07.268

Denkena, B., M.-A. Dittrich, F. Uhlich, and M. Wichmann. 2019. Approaches for an energy and resource efficient manufacturing in the aircraft industry. *Procedia CIRP* 80:180–185. https://doi.org/10.1016/j.procir.2019.01.101

Desbiolles, B.X.E., A. Bertsch, and P. Renaud. 2019. Ion beam etching redeposition for 3D multimaterial nanostructure manufacturing. *Microsystems and Nanoengineering* 5:11. https://doi.org/10.1038/s41378-019-0052-7

Dhakar, K., and A. Dvivedi. 2017. Dry and near-dry electric discharge machining processes. In: K. Gupta (Ed.), *Advanced Manufacturing Technologies. Materials Forming, Machining and Tribology*, pp. 249–266. Cham: Springer. https://doi.org/10.1007/978-3-319-56099-1_11

Di Maio, E., and E. Kiran. 2018. Foaming of polymers with supercritical fluids and perspectives on the current knowledge gaps and challenges. *The Journal of Supercritical Fluids* 134:157–166. https://doi.org/10.1016/j.supflu.2017.11.013

Dieterle, M., and T. Viere. 2021. Bridging product life cycle gaps in LCA & LCC towards a circular economy. *Procedia CIRP* 98:354–357. https://doi.org/10.1016/j.procir.2021.01.116

Dobrzański, L.A. 2013. *Description of Metals*. Gliwice: Silesian Technical University (in Polish).

Dom, R., G. Sivakumar, N.Y. Hebalkar, Sh.V. Joshi, and P.H. Borse. 2012. Deposition of nanostructured photocatalytic zinc ferrite films using solution precursor plasma spraying. *Materials Research Bulletin* 47(3):562–570. https://doi.org/10.1016/j.materresbull.2011.12.044

Dong, Y., W. Liu, H. Zhang, and H. Zhang. 2014. On-line recycling of abrasives in abrasive water jet cleaning. *Procedia CIRP* 15:278–282. https://doi.org/10.1016/j.procir.2014.06.045

Dorfman, M.R. 2018. Thermal spray coatings. In: M. Kutz (Ed.), *Handbook of Environmental Degradation of Materials* (Third Edition), pp. 469–488. Oxford, UK: William Andrew Publishing. https://doi.org/10.1016/B978-0-323-52472-8.00023-X

Doroshenko, V., V. Kravchenko, and O. Mul. 2012. Formation of the ice casting pattern structure and methods of its modeling. *Boundary Field Problems and Computer Simulation 51*(1):37–42.

Doroshenko, V.S. 2013. Gas-dynamic compaction of molding materials and a method for controlling the molding process. *The World of Technic and Technology* 2(135):60–66 (in Russian).

Doroshenko, V.S., and K.Kh. Berdyev. 2013. Gas-dynamic balance in sandy form during casting according to gasified models. *The World of Technic and Technology* 7(140):64–67 (in Russian).

Doroshenko, V.S., and O.I. Shinsk. 2009. Shell molding. *Metal and Foundry in Ukraine* 7–8:16–22 (in Russian).

Dragobetsky, V.V., A.D. Konovalenko, and V.G. Zagoryansky. 2012. *New and Highly Efficient Technologies in the Mechanical Engineering*. Kharkov: Tochka (in Russian).

Drobny, J.G. 2013. *Ionizing Radiation and Polymers*. Oxford, UK: William Andrew Publishing. https://doi.org/10.1016/B978-1-4557-7881-2.00006-7

Dryakhlov, E. 2009. Casting under pressure with foaming. *The World of Technic and Technology* 9(94):57–60 (in Russian).

Du, Ch., Sh. Chao, X. Gong, W. Ting, and X. Wei. 2018. Plasma methods for metals recovery from metal–containing waste. *Waste Management* 77:373–387. https://doi.org/10.1016/j.wasman.2018.04.026

Duan, W., X. Mei, Zh. Fan, J. Li, K. Wang, and Y. Zhang. 2020. Electrochemical corrosion assisted laser drilling of micro-hole without recast layer. *Optik* 202:163577. https://doi.org/10.1016/j.ijleo.2019.163577

Dudina, D.V., R.S. Mishra, and A.K. Mukherjee. 2016. Superplasticity. In: *Reference Module in Materials Science and Materials Engineering*. Amsterdam: Elsevier. https://doi.org/10.1016/B978-0-12-803581-8.02886-1

Dutta, N.K. 2003. Basic principles of laser diodes. In: C. Webb, and J. Jones (Eds.), *Handbook of Laser Technology and Applications*. Vol. 2, pp. 525–560. Boca Raton, FL: CRC Press. https://doi.org/10.1201/NOE0750309608

Ebnesajjad, A. 2003. *Melt Processible Fluoroplastics*. Norwich, NY: Plastics Design Library.

Endo, M., and R.F. Walter (Eds.). 2006. *Gas Lasers*. Boca Raton, FL: CRC Press.

Fahrenholtz, W.G., and G.E. Hilmas. 2017. Ultra-high temperature ceramics: Materials for extreme environments. *Scripta Materialia* 129:94–99. https://doi.org/10.1016/j.scriptamat.2016.10.018

Fan, Z., W. Jiang, F. Liu, and B. Xiao. 2014. Status quo and development trend of lost foam casting technology. *China Foundry* 11(4):296–307.

Fang F., and Z.W. Xu. 2018. Ion beam machining. In: S. Chatti, L. Laperrière, G. Reinhart, and T. Tolio (Eds.), *CIRP Encyclopedia of Production Engineering*. Berlin – Heidelberg: Springer. https://doi.org/10.1007/978-3-642-35950-7_6485-4

Fauchais, P.L., J.V.R. Heberlein, and M.I. Boulos. 2014. *Thermal Spray Fundamentals, from Powder to Part*. Berlin – Heidelberg: Springer. https://doi.org/10.1007/978-0-387-68991-3

Fegade, V., R.L. Shrivatsava, and A.V. Kale. 2015. Design for remanufacturing: Methods and their approaches. *Materials Today: Proceedings* 2(4–5):1849–1858. https://doi.org/10.1016/j.matpr.2015.07.130

Feldman, D. 2010. Polymeric foam materials for insulation in buildings. In: M.R. Hall (Ed.), *Materials for Energy Efficiency and Thermal Comfort in Buildings*, pp. 257–273. Cambridge, UK: Woodhead Publishing. https://doi.org/10.1533/9781845699277.2.257

Ferrara, M., and M. Bengisu. 2014. *Materials that Change Color*. Cham: Springer. https://doi.org/10.1007/978-3-319-00290-3_2

Filimon, A. 2019. Impact of smart structures on daily life. In: A. Filimon (Ed.), *Smart Materials: Integrated Design, Engineering Approaches, and Potential Applications*, pp. 1–10. Waretown, NJ: Apple Academic Press.

Fox-Rabinovich, G. 2013. Adaptive hard coatings design based on the concept of self-organization during friction. In: Q.J. Wang, and YW. Chung (Eds.), *Encyclopedia of Tribology*. Boston, MA: Springer. https://doi.org/10.1007/978-0-387-92897-5_1180

Francis, L.F. 2016. *Materials Processing*. San Diego, CA: Academic Press. https://doi.org/10.1016/B978-0-12-385132-1.00009-4

Fu, L., Q. Shi, Y. Ji, G. Wang, X. Zhang, J. Chen, Ch. Shen, and Ch.B. Park. 2020. Improved cell nucleating effect of partially melted crystal structure to enhance the microcellular foaming and impact properties of isotactic polypropylene. *The Journal of Supercritical Fluids* 160:104794. https://doi.org/10.1016/j.supflu.2020.104794

Galati, F., and B. Bigliardi. 2019. Industry 4.0: Emerging themes and future research avenues using a text mining approach. *Computers in Industry* 109:100–113. https://doi.org/10.1016/j.compind.2019.04.018

Gan, J.A., and C.C. Berndt. 2015. Plasma surface modification of metallic biomaterials. In: C. Wen (Ed.), *Surface Coating and Modification of Metallic Biomaterials*, pp. 103–157, Cambridge, UK: Woodhead Publishing. https://doi.org/10.1016/B978-1-78242-303-4.00004-1

Gangil, N., A.N. Siddiquee, and S. Maheshwari. 2021. *Composite Fabrication on Age-Hardened Alloy using Friction Stir Processing*. Boca Raton, London: CRC Press.

Gani, A., W. Ion, and E. Yang. 2021. Experimental investigation of plasma cutting two separate thin steel sheets simultaneously and parameters optimisation using Taguchi approach. *Journal of Manufacturing Processes* 64:1013–1023. https://doi.org/10.1016/j.jmapro.2021.01.055

Ganvir, A., N. Curry, N. Markocsan, P. Nylén, and F.-L. Toma. 2015. Comparative study of suspension plasma sprayed and suspension high velocity oxy-fuel sprayed YSZ thermal barrier coatings. *Surface and Coatings Technology* 268:70–76.

Ganvir, A., S. Goel, S. Govindarajan, A.R. Jahagirdar, S. Björklund, U. Klement, and Sh. Joshi. 2021. Tribological performance assessment of Al_2O_3-YSZ composite coatings deposited by hybrid powder-suspension plasma spraying. *Surface and Coatings Technology* 409:126907. https://doi.org/10.1016/j.surfcoat.2021.126907

Gao, S., and H. Huang. 2017. Recent advances in micro- and nano-machining technologies. *Frontiers of Mechanical Engineering* 12:18–32. https://doi.org/10.1007/s11465-017-0410-9

Garashchenko, Y., and M. Rucki. 2020. Part decomposition efficiency expectation evaluation in additive manufacturing process planning. *International Journal of Production Research* Article in Press. https://doi.org/10.1080/00207543.2020.1824084

Garrido, B., S. Dosta, and I.G. Cano. 2021. Bioactive glass coatings obtained by thermal spray: Current status and future challenges. *Boletín de la Sociedad Española de Cerámica y Vidrio* Article in Press. https://doi.org/10.1016/j.bsecv.2021.04.001

Gasch, M.J., D.T. Ellerby, and S.M. Johnson. 2005. Ultra high temperature ceramic composites. In: N.P. Bansal (Ed.), *Handbook of Ceramic Composites*. Boston, MA: Springer. https://doi.org/10.1007/0-387-23986-3_9

Gascoin, N., Ph. Gillard, and G. Baudry. 2009. Characterisation of oxidised aluminium powder: Validation of a new anodic oxidation bench. *Journal of Hazardous Materials* 171(1–3):348–357. https://doi.org/10.1016/j.jhazmat.2009.06.010

Gefvert, B., J. Wallace, G. Overton, and A. Nogee. 2019. Annual laser market review & forecast 2019. *Laser Focus World*. https://www.laserfocusworld.com/lasers-sources/article/16556290/annual-laser-market-review-forecast-2019-what-goes-up (accessed on February, 25, 2021).

Gellermann, W. 1991. Color center lasers. *Journal of Physics and Chemistry of Solids* 52(1):249–297. https://doi.org/10.1016/0022-3697(91)90068-B

Gerasimov, V.V., and O.V. Spirina. 2004. Low-melting borosilicate glazes for special-purpose and construction ceramics (a review). *Glass and Ceramics* 61(11–12):382–388.

Gevorkyan, E.S., and V.P. Nerubatskyi. 2009. Modeling of the Al_2O_3 hot pressing process with direct transmission of an alternating electric current with a frequency of 50 Hz. *Collection of Scientific Works of the Ukrainian State Academy of Railway Transport* 110:45–52.

Gevorkyan, E.S., V.P. Nerubatskyi, and O.M. Melnyk. 2010. Hot pressing of nanopowders of the composition ZrO_2–5%Y_2O_3. *Collection of Scientific Works of the Ukrainian State Academy of Railway Transport* 119:106–110.

Gevorkyan, E.S., L.A. Timofeeva, V.P. Nerubatskyi, and O.M. Melnyk. 2016. *Integrated Materials Processing Technologies: A Textbook*. Kharkiv: Ukrainian State University of Railway Transport (in Russian).

Ghisellini, P., and S. Ulgiati. 2020. Circular economy transition in Italy. Achievements, perspectives and constraints. *Journal of Cleaner Production* 243:118360. https://doi.org/10.1016/j.jclepro.2019.118360

Gibson, I., D. Rosen, B. Stucker, and M. Khorasani. 2021. *Additive Manufacturing Technologies*. Cham: Springer.

Gnanavelbabu, A., K.T. Sunu Surendran, P. Loganathan, and E. Vinothkumar. 2021. Effect of ageing temperature on the corrosion behaviour of UHTC particulates reinforced magnesium composites fabricated through ultrasonic assisted squeeze casting process. *Journal of Alloys and Compounds* 856:158173. https://doi.org/10.1016/j.jallcom.2020.158173

Gómez-Monterde, J., M. Schulte, S. Ilijevic, J. Hain, D. Arencón, M. Sánchez-Soto, M., and Ll. Maspoch. 2015. Morphology and mechanical characterization of ABS foamed by microcellular injection molding. *Procedia Engineering* 132:15–22. https://doi.org/10.1016/j.proeng.2015.12.462

Gómez-Monterde, J., J. Hain, M. Sánchez-Soto, and M. Ll. Maspoch. 2019. Microcellular injection moulding: A comparison between MuCell process and the novel micro-foaming technology IQ Foam. *Journal of Materials Processing Technology* 268:162–170. https://doi.org/10.1016/j.jmatprotec.2019.01.015

Goodridge, R., and S. Ziegelmeier. 2017. Powder bed fusion of polymers. In: M. Brandt (Ed.), *Laser Additive Manufacturing*, pp. 181–204. Duxford, UK: Woodhead Publishing. https://doi.org/10.1016/B978-0-08-100433-3.00007-5

Govindan, P., and V. Praveen. 2014. Developments in electrical discharge grinding process: A review. *International Journal of Engineering Sciences & Research Technology* 3(4):1932–1938.

Grebenok, T.P., T.V. Dubovik, M.S. Kovalchenko, L.A. Klochkov, A.A. Rogozinskaya, and V.I. Subbotin. 2016. Structure and properties of titanium carbide based cermets with additives of other carbides. *Powder Metallurgy and Metal Ceramics* 55:48–53. https://doi.org/10.1007/s11106-016-9779-y

Greer, A.L. 2014. Metallic glasses. In: D.E. Laughlin, K. Hono (Eds.), *Physical Metallurgy* (Fifth Edition), pp. 305–385. Amsterdam: Elsevier. https://doi.org/10.1016/B978-0-444-53770-6.00004-6

Guo, Zh., H. Liu, W. Dai, and Y. Lei. 2020. Responsive principles and applications of smart materials in biosensing. *Smart Materials in Medicine* 1:54–65. https://doi.org/10.1016/j.smaim.2020.07.001

Gusev, V.G., E.N. Petuhov, and A.M. Vukolov. 2012. The analysis of the current of the hydraulic medium through the nozzle of hydrocarved installation in program ANSYS. *Izvestia TulGU* 1:260–266 (in Russian).

Gushcha, A. 2009. Vacuum spraying. *The World of Technic and Technology* 10(95):50–51 (in Russian).

Gutsalenko, Yu.G., and A.V. Rudnev. 2020. Diamond spark grinding in anhydrated medium using solid lubricants. In: *Prospects and Priorities of Research in Science and Technology: Collective Monograph*. Vol. 1, pp. 44–59. Riga: Baltija Publishing. https://doi.org/10.30525/978-9934-26-008-7.1-3

Habrat, W.F. 2016. Effect of bond type and process parameters on grinding force components in grinding of cemented carbide. *Procedia Engineering* 149:122–129. https://doi.org/10.1016/j.proeng.2016.06.646

Halada, G.P., and C.R. Clayton. 2012. The intersection of design, manufacturing, and surface engineering. In: M. Kutz (Ed.), *Handbook of Environmental Degradation of Materials* (Second Edition), pp. 443–480. Oxford, UK: William Andrew Publishing. https://doi.org/10.1016/B978-1-4377-3455-3.00015-8

Han, L., M.M. Sartin, Zh.Q. Tian, D. Zhan, and Z.W. Tian. 2020. Electrochemical nanomachining. *Current Opinion in Electrochemistry* 22:80–86. https://doi.org/10.1016/j.coelec.2020.05.007

Haranzhevsky, E.V., and M.D. Krivilev. 2011. *Physics of the Lasers, Laser Technologies and Mathematical Modeling of the Laser Effect on the Substance*. Izhevsk: Udmurt State University.

Harfouche, A.L., D.A. Jacobson, D. Kainer, J.C. Romero, A.H. Harfouche, G.S. Mugnozza, M. Moshelion, G.A. Tuskan, J.J.B. Keurentjes, and A. Altman. 2019. Accelerating climate resilient plant breeding by applying next-generation artificial intelligence. *Trends in Biotechnology* 37(11):1217–1235. https://doi.org/10.1016/j.tibtech.2019.05.007

Harwani, D., V. Badheka, V. Patel, W. Li, and J. Andersson. 2021. Developing superplasticity in magnesium alloys with the help of friction stir processing and its variants: A review. *Journal of Materials Research and Technology* 12:2055–2075. https://doi.org/10.1016/j.jmrt.2021.03.115

Heimann, R.B. 2008. *Plasma Spray Coatings, Principles and Applications* (Second Edition). Oxford, UK: Wiley.

Henao, J., C. Poblano-Salas, M. Monsalve, J. Corona-Castuera, and O. Barceinas-Sanchez. 2019. Bio-active glass coatings manufactured by thermal spray: A status report. *Journal of Materials Research and Technology* 8(5):4965–4984. https://doi.org/10.1016/j.jmrt.2019.07.011

Hindmarch, A.T., D.E. Parkes, and A.W. Rushforth. 2012. Fabrication of metallic magnetic nanostructures by argon ion milling using a reversed-polarity planar magnetron ion source. *Vacuum* 86(10):1600–1604. https://doi.org/10.1016/j.vacuum.2012.02.019

Hitz, C.B., J.J. Ewing, and J. Hecht. 2001. *Introduction to Laser Technology*. Piscataway, NJ: IEEE Press Marketing.

Höche, D., J. Kaspar, and P. Schaaf. 2015. Laser nitriding and carburization of materials. In: J. Lawrence, and D.G. Waugh (Eds.), *Laser Surface Engineering*, pp. 33–58. Cambridge, UK: Woodhead Publishing. https://doi.org/10.1016/B978-1-78242-074-3.00002-7

Hocheng, H., H.Y. Tsai, U.U. Jadhav, K.Y. Wang, and T.C. Lin. 2014. Laser surface patterning. In: S. Hashmi, G. Ferreira Batalha, Ch.J. Van Tyne, B. Yilbas (Eds.), *Comprehensive Materials Processing*. Vol. 9, pp. 75–113. Amsterdam: Elsevier. https://doi.org/10.1016/B978-0-08-096532-1.00917-1

Höfer, R. 2012. Processing and performance additives for plastics. In: K. Matyjaszewski, and M. Möller (Eds.), *Polymer Science: A Comprehensive Reference*, pp. 369–381. Amsterdam: Elsevier. https://doi.org/10.1016/B978-0-444-53349-4.00272-7

Hopkins, C., and A. Hosseini. 2019. A review of developments in the fields of the design of smart cutting tools, wear monitoring, and sensor innovation. *IFAC-PapersOnLine* 52(10):352–357. https://doi.org/10.1016/j.ifacol.2019.10.056

Hossain, M.A., A.N. Ahmed, and M.A.A. Khan. 2014. Vitreous enamel coating on mild steel substrate: Characterization and evaluation. *International Journal of Scientific & Engineering Research* 5(2):821–826.

Hou, H., J. Veilleux, F. Gitzhofer, Q. Wang, and Y. Liu. 2019. Hybrid suspension/solution precursor plasma spraying of a complex Ba(Mg1/3Ta2/3)O3 Perovskite: Effects of processing parameters and precursor chemistry on phase formation and decomposition. *Journal of Thermal Spray Technology* 28:12–26. https://doi.org/10.1007/s11666-018-0797-9

Hourmand, M., A.A.D. Sarhan, M.Y. Noordin, and M. Sayuti. 2017. Micro-EDM drilling of tungsten carbide using microelectrode with high aspect ratio to improve MRR, EWR, and hole quality. In: M.S.J. Hashmi (Ed.), *Comprehensive Materials Finishing*. Vol. 1, pp. 267–321. Amsterdam: Elsevier. https://doi.org/10.1016/B978-0-12-803581-8.09155-4

Hu, J.L. 2016. Introduction to active coatings for smart textiles. In: J.L. Hu (Ed.), *Active Coatings for Smart Textiles*, pp. 1–7. Duxford, UK: Woodhead Publishing. https://doi.org/10.1016/B978-0-08-100263-6.00001-0

Hu, Z., R. Myong, A. Nguyen, Z. Jiang, and T. Cho. 2007. Numerical analysis of the flowfield in a supersonic coil with an interleaved jet configuration and its effect on the gain distribution. *Engineering Applications of Computational Fluid Mechanics* 1(3):207–215.

Huang, G., Y. Wang, M. Zhang, Ch. Cui, and Zh. Tong. 2021. Brazing diamond grits onto AA7075 aluminium alloy substrate with Ag–Cu–Ti filler alloy by laser heating. *Chinese Journal of Aeronautics* 34(6):67–78. https://doi.org/10.1016/j.cja.2020.07.005

Huang, J., Q. Chang, J. Arinez, and G. Xiao. 2019. A maintenance and energy saving joint control scheme for sustainable manufacturing systems. *Procedia CIRP* 80:263–268. https://doi.org/10.1016/j.procir.2019.01.073

Humphreys, J., G.S. Rohrer, and A. Rollett. 2017. *Recrystallization and related annealing phenomena* (Third Edition). Amsterdam: Elsevier. https://doi.org/10.1016/B978-0-08-098235-9.00015-X

ISO. 23472-1:2020. *Foundry Machinery—Terminology—Part 1: Fundamental Terminology.*

Ivanov, A., R. Leese, and A. Spieser. 2015. Micro-electrochemical machining. In: Y. Qin (Ed.), *Micro and Nano Technologies, Micromanufacturing Engineering and Technology* (Second Edition), pp. 121–145. Oxford, UK: William Andrew Publishing. https://doi.org/10.1016/B978-0-323-31149-6.00006-2

Ivanova, T.N., A.I. Korshunov, and P. Bozek. 2018. The influence of chemical composition of tough-to-machine steels on grinding technologies. *Management Systems in Production Engineering* 26(3):172–177. https://doi.org/10.1515/mspe-2018-0028

Izwan, N.Sh.L.B., Zh. Feng, J.B. Patel, and W.N. Hung. 2016. Prediction of material removal rate in die-sinking electrical discharge machining. *Procedia Manufacturing* 5:658–668. https://doi.org/10.1016/j.promfg.2016.08.054

Jain, N.K., and S. Pathak. 2017. Electrochemical processing and surface finish. In: M.S.J. Hashmi (Ed.), *Comprehensive Materials Finishing*. Vol. 1, pp. 358–380. Amsterdam: Elsevier. https://doi.org/10.1016/B978-0-12-803581-8.09182-7

Jain, V., A.K. Sharma, and P. Kumar. 2011. Recent developments and research issues in microultrasonic machining. *International Scholarly Research Notices* 2011:413231. https://doi.org/10.5402/2011/413231

Jain, V.K., A. Sidpara, M. Ravisankar, and M. Das. 2013. Micromanufacturing: An introduction. In: V.K. Jain (Ed.), *Micromanufacturing Processes*, pp. 3–38. Boca Raton, FL: CRC Press. https://doi.org/10.1201/b13020

Jawahir, I.S., E. Brinksmeier, R. M'Saoubi, D.K. Aspinwall, J.C. Outeiro, D. Meyer, D. Umbrello, and A.D. Jayal. 2011. Surface integrity in material removal processes: Recent advances. *CIRP Annals* 60(2):603–626. https://doi.org/10.1016/j.cirp.2011.05.002

Jhavar, S., C.P. Paul, and N.K. Jain. 2016. Micro-plasma transferred arc additive manufacturing for die and mold surface remanufacturing. *JOM* 68:1801–1809. https://doi.org/10.1007/s11837-016-1932-z

Jia, Zh., J. Ma, D. Song, F. Wang, and W. Liu. 2018. A review of contouring-error reduction method in multi-axis CNC machining. *International Journal of Machine Tools and Manufacture* 125:34–54. https://doi.org/10.1016/j.ijmachtools.2017.10.008

Jiang, W., and Z. Fan. 2018. Novel technologies for the lost foam casting process. *Frontiers of Mechenical Engineering* 13:37–47. https://doi.org/10.1007/s11465-018-0473-2

Jiang, N., R. Xie, Q. Liu, and J. Li. 2017. Fabrication of sub-micrometer MgO transparent ceramics by spark plasma sintering. *Journal of the European Ceramic Society* 37(15):4947–4953. https://doi.org/10.1016/j.jeurceramsoc.2017.06.021

Jimoda, L.A. 2011. An environmental audit of methylene chloride blown batch and continuous polyurethane foam production process: Emission qualitative distribution, Auxilliary Blowing Agent tunnel concentration and diffusion coefficient estimations. FACTA UNIVERSITATIS Series: Working and Living Environmental Protection 8(1):51–62.

Jin, F.L., M. Zhao, M. Park, and S.J. Park. 2019. Recent trends of foaming in polymer processing: A review. *Polymers* 11(6):953. https://doi.org/10.3390/polym11060953

Joshi, A.Y., and A.Y. Joshi. 2019. A systematic review on powder mixed electrical discharge machining. *Heliyon* 5(12):e02963. https://doi.org/10.1016/j.heliyon.2019.e02963

Joshi, S.V., G. Sivakumar, T. Raghuveer, and R.O. Dusane. 2014. Hybrid plasma-sprayed thermal barrier coatings using powder and solution precursor feedstock. *Journal of Thermal Spray Technology* 23:616–624. https://doi.org/10.1007/s11666-014-0075-4

Kafle, B.P. 2020. *Chemical Analysis and Material Characterization by Spectrophotometry*. Amsterdam: Elsevier. https://doi.org/10.1016/B978-0-12-814866-2.00006-3

Kafuku, J.M., M.Z.M. Saman, and S.M. Yusof. 2019. Application of fuzzy logic in selection of remanufacturing technology. *Procedia Manufacturing* 33:192–199. https://doi.org/10.1016/j.promfg.2019.04.023

Kaiser, A., M. Lobert, and R. Telle. 2008. Thermal stability of zircon ($ZrSiO_4$). *Journal of the European Ceramic Society* 28:2199–2211. https://doi.org/10.1016/j.jeurceramsoc.2007.12.040

Kalaiselvan, K., I. Dinaharan, and N. Murugan. 2021. Routes for the joining of metal matrix composite materials. *Reference Module in Materials Science and Materials Engineering*. Amsterdam: Elsevier. https://doi.org/10.1016/B978-0-12-803581-8.11899-5

Kang, I.S., J.S. Kim, Y.W. Seo, and J.H. Kim. 2006. An experimental study on the ultrasonic machining characteristics of engineering ceramics. *Journal of Mechanical Science and Technology* 20:227–233. https://doi.org/10.1007/BF02915824

Kanyilmaz, A. 2019. The problematic nature of steel hollow section joint fabrication, and a remedy using laser cutting technology: A review of research, applications, opportunities. *Engineering Structures* 183:1027–1048. https://doi.org/10.1016/j.engstruct.2018.12.080

Kapoor, R. 2017. Severe plastic deformation of materials. In: A.K. Tyagi, and S. Banerjee (Eds.), *Materials Under Extreme Conditions*, pp. 717–754. Amsterdam: Elsevier. https://doi.org/10.1016/B978-0-12-801300-7.00020-6

Karmakar, B. 2016. Fundamentals of glass and glass nanocomposites. In: B. Karmakar, K. Rademann, and A.L. Stepanov (Eds.), *Glass Nanocomposites*, pp. 3–53. Oxford, UK: William Andrew Publishing. https://doi.org/10.1016/B978-0-323-39309-6.00001-8

Kassner, H., R. Siegert, D. Hathiramani, R. Vassen, and D. Stoever. 2008. Application of suspension plasma spraying (SPS) for manufacture of ceramic coatings. *Journal of Thermal Spray Technology* 17:115–123. https://doi.org/10.1007/s11666-007-9144-2

Kastner, C., and G. Steinbichler. 2020. Theoretical background and automation approach for a novel measurement method for determining dynamic solubility limits of supercritical fluids in injection foam molding. *Polymer Engineering and Science* 60(2):330–340. https://doi.org/10.1002/pen.25288

Kastner, C., T. Mitterlehner, D. Altmann, and G. Steinbichler. 2020. Backpressure optimization in foam injection molding: Method and assessment of sustainability. *Polymers* 12(11):2696. https://doi.org/10.3390/polym12112696

Katayama, S. 2018. Understanding and improving process control in pulsed and continuous wave laser welding. In: J. Lawrence (Ed.), *Advances in Laser Materials Processing* (Second Edition), pp. 153–183. Cambridge, MA: Woodhead Publishing. https://doi.org/10.1016/B978-0-08-101252-9.00007-8

Katnam, K.B., L.F.M. Da Silva, and T.M. Young. 2013. Bonded repair of composite aircraft structures: A review of scientific challenges and opportunities. *Progress in Aerospace Sciences* 61:26–42. https://doi.org/10.1016/j.paerosci.2013.03.003

Katsamas, A.I., and G.N. Haidemenopoulos. 2001. Laser-beam carburizing of low-alloy steels. *Surface and Coatings Technology* 139(2–3):183–191. https://doi.org/10.1016/S0257-8972(00)01061-6

Kaufman, C.M., and M.R. Overcash. 1993. Waste minimization in the manufacture of flexible polyurethane foams: Quantification of auxiliary blowing agent volatilization. *Air & Waste* 43(5):736–744. https://doi.org/10.1080/1073161X.1993.10467155

Kazak, A.K., and V.V. Didenko. 2016. Development of high-temperature nanostructured silicate-enamel coatings for enamelling engineering metalware. *Metallurgist* 59:1125–1128. https://doi.org/10.1007/s11015-016-0226-6

Kazarian S.G. 2000. Polymer processing with supercritical fluids. *Polymer Science, Series C* 42(1):78–101.

Kerin, M., and D.T. Pham. 2019. A review of emerging industry 4.0 technologies in remanufacturing. *Journal of Cleaner Production* 237:117805. https://doi.org/10.1016/j.jclepro.2019.117805

Kim, B.N., K., Hiraga, K. Morita, and Y. Sakka. 2001. A high-strain-rate superplastic ceramic. *Nature* 413:288–291. https://doi.org/10.1038/35095025

Kim, H.S., B.R. Kang, and S.M. 2020. Choi microstructure and mechanical properties of vacuum plasma sprayed HfC, TiC, and HfC/TiC ultra-high-temperature ceramic coatings. *Materials* 13(1):124. https://doi.org/10.3390/ma13010124

Kim, K.W., G.-S. Ham, G.-S. Cho, Ch.P. Kim, S.-Ch. Park, and K.-A. Lee. 2021. Microstructures and corrosion properties of novel Fe46.8-Mo30.6-Cr16.6-C4.3-B1.7 metallic glass coatings manufactured by vacuum plasma spray process. *Intermetallics* 130:107061. https://doi.org/10.1016/j.intermet.2020.107061

Kmetty, Á., K. Litauszki, and D. Réti. 2018. Characterization of different chemical blowing agents and their applicability to produce poly(lactic acid) foams by extrusion. *Applied Sciences* 8(10):1960. https://doi.org/10.3390/app8101960

Koch, S.W., and M.R. Hofmann. 2018. Semiconductor lasers. In: B.D. Guenther, and D.G. Steel (Eds.), *Encyclopedia of Modern Optics* (Second Edition), pp. 462–468. Amsterdam: Elsevier. https://doi.org/10.1016/B978-0-12-803581-8.10103-1

König, W., L. Cronjäger, G. Spur, H.K. Tönshoff, M. Vigneau, and W.J. Zdeblick. 1990. Machining of new materials. *CIRP Annals* 39(2):673–681. https://doi.org/10.1016/S0007-8506(07)63004-2

Koryagin, S.I., I.V. Pimenov, and V.K. Khudyakov. 2000. *Materials Processing Methods*. Kaliningrad: University of Kaliningrad (in Russian).

Korzhov, E.G. 2006. Some peculiarities of the waterjet machinning of the materials. *Mining Informational and Analytical Bulletin* 3:273–287 (in Russian).

Kumar, K., D. Zindani, and J.P. Davim. 2018. *Advanced Machining and Manufacturing Processes*. Cham: Springer. https://doi.org/10.1007/978-3-319-76075-9

Kumar, M.Bh., and P. Sathiya. 2021. Methods and materials for additive manufacturing: A critical review on advancements and challenges. *Thin-Walled Structures* 159:107228. https://doi.org/10.1016/j.tws.2020.107228

Kundas, S., and A. Ilyuschenko. 2002. Computer simulation and control of plasma spraying processes. *Materials and Manufacturing Processes* 17(1):85–96. https://doi.org/10.1081/AMP-120002799

Lahrour, Y., and D. Brissaud. 2018. A technical assessment of product/component re-manufacturability for additive remanufacturing. *Procedia CIRP* 69:142–147. https://doi.org/10.1016/j.procir.2017.11.105

Langdon, T.G. 1982. The mechanical properties of superplastic materials. *Metallurgical and Materials Transactions* 13:689–701. https://doi.org/10.1007/BF02642383

Laser Marking Market. 2021. https://www.marketsandmarkets.com/Market-Reports/laser-marking-market-167085735.html (accessed on February, 24, 2021).

Łatka, L., L. Pawłowski, M. Winnicki, P. Sokołowski, A. Małachowska, and S. Kozerski. 2020. Review of functionally graded thermal sprayed coatings. *Applied Sciences* 10(15):5153. https://doi.org/10.3390/app10155153

Lauwers, B., J. Vleugels, O. Malek, K. Brans, and K. Liu. 2012. Electrical discharge machining of composites. In: H. Hocheng (Ed.), *Machining Technology for Composite Materials*, pp. 202–241. Cambridge, UK: Woodhead Publishing. https://doi.org/10.1533/9780857095145.2.202

Lavrentiev, A.Yu., and M.A. Lavrentiev. 2015. Protection of the welded constructions from oxidation during heat treatment. *Science, Technology and Education* 11(17):28–31 (in Russian).

Lazareva, E.A. 2020. Heat-resistant sitall coatings for high-temperature corrosion protection of nichrome alloys. *Materials Science Forum* 992:633–639. https://doi.org/10.4028/www.scientific.net/msf.992.633

Lazorenko, G., A. Kasprzhitskii, and T. Nazdracheva. 2021. Anti-corrosion coatings for protection of steel railway structures exposed to atmospheric environments: A review. *Construction and Building Materials* 288:123115. https://doi.org/10.1016/j.conbuildmat.2021.123115

Le, V.Th., H. Paris, and G. Mandil. 2015. Using additive and subtractive manufacturing technologies in a new remanufacturing strategy to produce new parts from end-of-life parts. In: Congrès Français de Mécanique 2015, Association Française de Mécanique (AFM), August 2015, Lyon, France.

Le, V.Th., H. Paris, and G. Mandil. 2017. Process planning for combined additive and subtractive manufacturing technologies in a remanufacturing context. *Journal of Manufacturing Systems* 44(Part 1):243–254. https://doi.org/10.1016/j.jmsy.2017.06.003

Leese, R.J., and A. Ivanov. 2016. Electrochemical micromachining: An introduction. *Advances in Mechanical Engineering* 8(1):1–13. https://doi.org/10.1177/1687814015626860

Lei, J., Q. Shengru, Zh. Chengyu, M. Kan, H. Dong, D. Gertsriken, and V. Mazanko. 2010. Fatigue properties of Ti17 alloy strengthened by combination of electric spark treatment with ultrasonic surface treatment. *Rare Metal Materials and Engineering* 39(12):2091–2094. https://doi.org/10.1016/S1875-5372(11)60006-4

Łępicka, M., M. Grądzka-Dahlke, D. Pieniak, K. Pasierbiewicz, K. Kryńska, and A. Niewczas. 2019. Tribological performance of titanium nitride coatings: A comparative study on TiN-coated stainless steel and titanium alloy. *Wear* 422–423:68–80. https://doi.org/10.1016/j.wear.2019.01.029

Li, C.J. 2010. Thermal spraying of light alloys. In: H. Dong (Ed.), *Surface Engineering of Light Alloys*, pp. 184–241. Cambridge, UK: Woodhead Publishing. https://doi.org/10.1533/9781845699451.2.184

Li, Z., Z. Huang, F. Sun, X. Li, and J. Ma. 2020. Forming of metallic glasses: Mechanisms and processes. *Materials Today Advances* 7:100077. https://doi.org/10.1016/j.mtadv.2020.100077

Li, Zh., B. Zhao, and W. Chen. 2015. Superplastic forming and diffusion bonding: Progress and trends. *MATEC Web of Conferences* 21:01005. https://doi.org/10.1051/matecconf/20152101005

Liang, S.Y., and A.J. Shih. 2016. *Analysis of Machining and Machine Tools*. New York: Springer.

Lima, C., F. Andrade, C. Kawakami, C. Gonçalves, and W. Peraro. 2016. Methods to improve the surface quality of microcellular injection molded parts: A review. *SAE Technical Paper* 36:0224. https://doi.org/10.4271/2016-36-0224

Liobikienė, G., and A. Minelgaitė. 2021. Energy and resource-saving behaviours in European Union countries: The Campbell paradigm and goal framing theory approaches. *Science of The Total Environment* 750:141745. https://doi.org/10.1016/j.scitotenv.2020.14174

Liu, C., W. Cai, Sh. Jia, M. Zhang, H. Guo, L. Hu, and Zh. Jiang. 2018. Emergy-based evaluation and improvement for sustainable manufacturing systems considering resource efficiency and environment performance. *Energy Conversion and Management* 177:176–189. https://doi.org/10.1016/j.enconman.2018.09.039

Liu, G., X. Zhang, X. Chen, Y. He, L. Cheng, M. Huo, J. Yin, F. Hao, S. Chen, P. Wang, Sh. Yi, L. Wan, Zh. Mao, Zh. Chen, X. Wang, Zh. Cao, and J. Lu. 2021. Additive manufacturing of structural materials. *Materials Science and Engineering: R: Reports*, 145:100596. https://doi.org/10.1016/j.mser.2020.100596

Liu, P.S., and G.F. Chen. 2014. *Porous Materials*. Oxford, UK: Elsevier. https://doi.org/10.1016/C2012-0-03669-1

Liu, Q., and M.C. Leu. 2006. Investigation of interface agent for investment casting with ice patterns. *Journal of Manufacturing Science and Engineering* 128(2):554–562. https://doi.org/10.1115/1.2162902

Liu, W.W., Z.J. Tang, X.Y. Liu, H.J. Wang, and H.Ch. Zhang. 2017. A review on in-situ monitoring and adaptive control technology for laser cladding remanufacturing. *Procedia CIRP* 61:235–240. https://doi.org/10.1016/j.procir.2016.11.217

Liu, X., Z. Liang, G. Wen, X. Yuan. 2019. Waterjet machining and research developments: A review. *The International Journal of Advanced Manufacturing Technology* 102:1257–1335. https://doi.org/10.1007/s00170-018-3094-3

Liu, Zh., Q. Jiang, T. Li, Sh. Dong, Sh. Yan, H. Zhang, and B. Xu. 2016. Environmental benefits of remanufacturing: A case study of cylinder heads remanufactured through laser cladding. *Journal of Cleaner Production* 133:1027–1033. https://doi.org/10.1016/j.jclepro.2016.06.049

Llanto, J.M., M. Tolouei-Rad, A. Vafadar, and M. Aamir. 2021. Recent progress trend on abrasive waterjet cutting of metallic materials: A review. *Applied Sciences* 11(8):3344. https://doi.org/10.3390/app11083344

Lohia, A., G. Sivakumar, M. Ramakrishna, and S.V. Joshi. 2014. Deposition of nanocomposite coatings employing a hybrid APS + SPPS technique. *Journal of Thermal Spray Technology* 23:1054–1064. https://doi.org/10.1007/s11666-014-0071-8

Lutey, A., A. Ascari, A. Fortunato, and L. Romoli. 2018. Long-pulse quasi-CW laser cutting of metals. *The International Journal of Advanced Manufacturing Technology* 94:155–162. https://doi.org/10.1007/s00170-017-0913-x

Łyczkowska-Widłak, E., P. Lochyński, and G. Nawrat. 2020. Electrochemical polishing of austenitic stainless steels. *Materials* 13:2557. https://doi.org/10.3390/ma13112557

Madhuri, K.V. 2020. Thermal protection coatings of metal oxide powders. In: Y. Al-Douri (Ed.), *Metal Oxide Powder Technologies*, pp. 209–231. Amsterdam: Elsevier. https://doi.org/10.1016/B978-0-12-817505-7.00010-5

Maggi, F., M. Balduzzi, R. Vosseler, M. Rösler, W. Quadrini, G. Tavola, M. Pogliani, D. Quarta, and S. Zanero. 2021. Smart factory security: A case study on a modular smart manufacturing system. *Procedia Computer Science* 180:666–675. https://doi.org/10.1016/j.procs.2021.01.289

Mahade, S., K Narayan, S Govindarajan, S Björklund, N Curry, and S. Joshi. 2019. Exploiting suspension plasma spraying to deposit wear-resistant carbide coatings. *Materials* 12(15):2344. https://doi.org/10.3390/ma12152344

Mahade, S., S. Björklund, S. Govindarajan, M. Olsson, and Sh. Joshi. 2020. Novel wear resistant carbide-laden coatings deposited by powder-suspension hybrid plasma spray: Characterization and testing. *Surface and Coatings Technology* 399:126147. https://doi.org/10.1016/j.surfcoat.2020.126147

Maitra, S. 2014. Nanoceramic matrix composites: types, processing and applications. In: I.M. Low (Ed.), *Advances in Ceramic Matrix Composites*, pp. 27–42. Cambridge, UK: Woodhead Publishing. https://doi.org/10.1533/9780857098825.1.27

Mamun, K.A., and J. Stokes. 2014. Development of a semi-automated dual feed unit to produce FGM coatings using the HVOF thermal spray process. South Pacific Journal of Natural Science. 32:18. https://doi.org/10.1071/SP14003

Martin, P.M. 2011. *Introduction to Surface Engineering and Functionally Engineered Materials*. Hoboken, NJ: Scrivener – Wiley.

Martinsen, K., S.J. Hu, and B.E. Carlson. 2015. Joining of dissimilar materials. *CIRP Annals* 64(2):679–699. https://doi.org/10.1016/j.cirp.2015.05.006

Masuda, H., and E. Sato. 2020. Diffusional and dislocation accommodation mechanisms in superplastic materials. *Acta Materialia* 197:235–252. https://doi.org/10.1016/j.actamat.2020.07.042

Mattox, D.M. 2010. *Handbook of Physical Vapor Deposition (PVD) Processing* (Second Edition). Oxford, UK: William Andrew Publishing. https://doi.org/10.1016/B978-0-8155-2037-5.00015-0

Mayrhofer, P.H., C. Mitterer, and H. Clemens. 2005. Self-organized nanostructures in hard ceramic coatings. *Advanced Engineering Materials* 7(12):1071–1082. https://doi.org/10.1002/adem.200500154

Mayrhofer, P.H., Ch. Mitterer, L. Hultman, and H. Clemens. 2006. Microstructural design of hard coatings. *Progress in Materials Science* 51(8):1032–1114. https://doi.org/10.1016/j.pmatsci.2006.02.002,

McLaren, D., J. Niskanen, and J. Anshelm. 2020. Reconfiguring repair: Contested politics and values of repair challenge instrumental discourses found in circular economies literature. *Resources, Conservation & Recycling: X* 8:100046. https://doi.org/10.1016/j.rcrx.2020.100046

Merklein, M., M. Johannes, M. Lechner, and A. Kuppert. 2014. A review on tailored blanks: Production, applications and evaluation. *Journal of Materials Processing Technology* 214(2):151–164. https://doi.org/10.1016/j.jmatprotec.2013.08.015

Merklein, M., D. Junker, A. Schaub, and F. Neubauer. 2016. Hybrid additive manufacturing technologies: An analysis regarding potentials and applications. *Physics Procedia* 83:549–559. https://doi.org/10.1016/j.phpro.2016.08.057

Meschut, G., V. Janzen, and T. Olfermann. 2014. Innovative and highly productive joining technologies for multi-material lightweight car body structures. *Journal of Materials Engineering and Performance* 23:1515–1523. https://doi.org/10.1007/s11665-014-0962-3

Messler, R.W. 2006. *Integral Mechanical Attachment*. Oxford, UK: Butterworth-Heinemann. https://doi.org/10.1016/B978-075067965-7/50027-4

Middeldorf, K., and D. von Hofe. 2008. Trends in the development of joining techniques. *Automatic Welding* 11:39–47 (in Russian).

Miorin, E., F. Montagner, V. Zin, D. Giuranno, E. Ricci, M. Pedroni, V. Spampinato, E. Vassallo, and S.M. Deambrosis. 2019. Al rich PVD protective coatings: A promising approach to prevent T91 steel corrosion in stagnant liquid lead. *Surface and Coatings Technology* 377:124890. https://doi.org/10.1016/j.surfcoat.2019.124890

Mitterer, Ch. 2014. PVD and CVD hard coatings. In: V.K. Sarin (Ed.), *Comprehensive Hard Materials*, pp. 449–467. Amsterdam: Elsevier. https://doi.org/10.1016/B978-0-08-096527-7.00035-0

Mohamed, F.A., and Y. Li. 2001. Creep and superplasticity in nanocrystalline materials: current understanding and future prospects. *Materials Science and Engineering: A* 298 (1–2):1–15. https://doi.org/10.1016/S0928-4931(00)00190-9

Mohammad, A.E.K., and D. Wang. 2016. Electrochemical mechanical polishing technology: recent developments and future research and industrial needs. *The International Journal of Advanced Manufacturing Technology* 86:1909–1924. https://doi.org/10.1007/s00170-015-8119-6

Möhring, H.-C., K. Werkle, and W. Maier. 2020. Process monitoring with a cyber-physical cutting tool. *Procedia CIRP* 93:1466–1471. https://doi.org/10.1016/j.procir.2020.03.034

Mondal, S. 2008. Phase change materials for smart textiles: An overview. *Applied Thermal Engineering* 28(11–12):1536–1550. https://doi.org/10.1016/j.applthermaleng.2007.08.009

Morks, M.F., I. Cole, and A. Kobayashi. 2013. Plasma forming multilayer ceramics for ultra-high temperature application. *Vacuum* 88:134–138. https://doi.org/10.1016/j.vacuum.2012.03.045

Morozow, D., J. Narojczyk, M. Rucki, and S. Lavrynenko. 2018. Wear resistance of the cermet cutting tools after aluminum (Al^+) and nitrogen (N^+) ion implantation. *Advances in Materials Science* 18(2):92–99. https://doi.org/10.1515/adms-2017-0035

Morozow, D., M. Rucki, Z. Siemiatkowski, and Yu. Gutsalenko. 2019. Tribological tests of the ceramic cutting tools after yttrium (Y^+) and rhenium (Re^+) ion implantation. *Tribologia* 285(3):71–77. https://doi.org/10.5604/01.3001.0013.5436

Morozow, D., Z. Siemiątkowski, E. Gevorkyan, M. Rucki, J. Matijošius, A. Kilikevičius, J. Caban, and Z. Krzysiak. 2020. Effect of yttrium and rhenium ion implantation on the performance of nitride ceramic cutting tools. *Materials* 13:4687. https://doi.org/10.3390/ma13204687

Morozow, D., M. Barlak, Z. Werner, M. Pisarek, P. Konarski, J. Zagórski, M. Rucki, L. Chałko, M. Łagodziński, J. Narojczyk, Z. Krzysiak, and J. Caban. 2021. Wear resistance improvement of cemented tungsten carbide deep-hole drills after ion implantation. *Materials* 14(2):239. https://doi.org/10.3390/ma14020239

Mostaghimi, J., L. Pershin, and S. Yugeswaran. 2017. Heat transfer in DC AND RF plasma torches. In: F. Kulacki (Ed.), *Handbook of Thermal Science and Engineering*. Cham: Springer. https://doi.org/10.1007/978-3-319-32003-8_27-1

Mouritz, A.P. 2012. *Introduction to Aerospace Materials*. Cambridge, UK: Woodhead Publishing. https://doi.org/10.1533/9780857095152.154

Mubarok, F., and N. Espallargas. 2015. Suspension plasma spraying of sub-micron silicon carbide composite coatings. *Journal of Thermal Spray Technology* 24:817–825. https://doi.org/10.1007/s11666-015-0242-2

Mukherjee, A., Deepmala, P. Srivastava, and J.K. Sandhu. 2021. Application of smart materials in civil engineering: A review. *Materials Today: Proceedings*, Article in Press. https://doi.org/10.1016/j.matpr.2021.03.304

Muresan, L.M. 2015. Corrosion protective coatings for Ti and Ti alloys used for biomedical implants. In: A. Tiwari, J. Rawlins, and Ll.H. Hihara (Eds.), *Intelligent Coatings for Corrosion Control*, pp. 585–602. Oxford, UK: Butterworth-Heinemann. https://doi.org/10.1016/B978-0-12-411467-8.00017-9

Nagalingam, S.V., and G.C.I. Lin. 2008. CIM: Still the solution for manufacturing industry. *Robotics and Computer-Integrated Manufacturing* 24(3):332–344. https://doi.org/10.1016/j.rcim.2007.01.002

Nagimova, A., and A. Perveen. 2019. A review on laser machining of hard to cut materials. *Materials Today: Proceedings* 18(Part 7):2440–2447. https://doi.org/10.1016/j.matpr.2019.07.092

Nalawade, S.P., F. Picchioni, and L.P.B.M. Janssen. 2006. Supercritical carbon dioxide as a green solvent for processing polymer melts: Processing aspects and applications. *Progress in Polymer Science* 31:19–43. https://doi.org/10.1016/j.progpolymsci.2005.08.002

Narojczyk, J., D. Morozow, J.W. Narojczyk, and M. Rucki. 2018. Ion implantation of the tool's rake face for machining of the Ti-6Al-4V alloy. *Journal of Manufacturing Processes* 34(Part A):274–280. https://doi.org/10.1016/j.jmapro.2018.06.017

Natarajan, Y., P.K. Murugesan, M. Mohan, and Sh.A.L. Ali Khan. 2020. Abrasive water jet machining process: A state of art of review. *Journal of Manufacturing Processes* 49:271–322. https://doi.org/10.1016/j.jmapro.2019.11.030

Nath, A.K. 2014. Laser drilling of metallic and nonmetallic substrates. In: S. Hashmi, G. Ferreira Batalha, Ch.J. Van Tyne, and B. Yilbas (Eds.), *Comprehensive Materials Processing*. Vol. 9, pp. 115–175. Amsterdam: Elsevier. https://doi.org/10.1016/B978-0-08-096532-1.00904-3

National Research Council. 1994. *Free Electron Lasers and Other Advanced Sources of Light: Scientific Research Opportunities*. Washington, DC: The National Academies Press. https://doi.org/10.17226/9182

Neikov, O.D. 2019. Powders for additive manufacturing processing. In: O.D. Neikov, S.S. Naboychenko, and N.A. Yefimov (Eds.), *Handbook of Non-Ferrous Metal Powders* (Second Edition), pp. 373–399. Amsterdam: Elsevier. https://doi.org/10.1016/B978-0-08-100543-9.00013-0

Ni, J., K. Yu, H. Zhou, J. Mi, Sh. Chen, and X. Wang. 2020. Morphological evolution of PLA foam from microcellular to nanocellular induced by cold crystallization assisted by supercritical CO_2. *The Journal of Supercritical Fluids* 158:104719. https://doi.org/10.1016/j.supflu.2019.104719

Niaounakis, M. 2015. *Biopolymers: Processing and Products.* Oxford, UK: William Andrew Publishing. https://doi.org/10.1016/B978-0-323-26698-7.00009-X

Nofar, M., and Ch.B. Park. 2018. *Polylactide Foams: Fundamentals, Manufacturing, and Applications.* Oxford, UK: Elsevier. https://doi.org/10.1016/C2017-0-00939-4

Noginov, M., G. Zhu, A. Belgrave, R. Bakker, V.M. Shalaev, E.E. Narimanov, S. Stout, E. Herz, T. Suteewong, and U. Wiesner. 2009. Demonstration of a spaser-based nanolaser. *Nature* 460:1110–1112. https://doi.org/10.1038/nature08318

Oh, S.Y., J.S. Shin, S. Park, T.S. Kim, H. Park, L. Lee, and J. Lee. 2021. Underwater laser cutting of thick stainless steel blocks using single and dual nozzles. *Optics & Laser Technology* 136:106757. https://doi.org/10.1016/j.optlastec.2020.106757

Okada, A. 2019. Electron beam machining. In: S. Chatti, L. Laperrière, G. Reinhart, and T. Tolio (Eds.), *CIRP Encyclopedia of Production Engineering.* Berlin – Heidelberg: Springer. https://doi.org/10.1007/978-3-662-53120-4_6480

Okolieocha, Ch., D. Raps, K. Subramaniam, and V. Altstädt. 2015. Microcellular to nanocellular polymer foams: Progress (2004–2015) and future directions: A review. *European Polymer Journal* 73:500–519. https://doi.org/10.1016/j.eurpolymj.2015.11.001

Orbanic, H., and M. Junkar. 2008. Analysis of striation formation mechanism in abrasive water jet cutting. *Wear* 265(5–6):821–830. https://doi.org/10.1016/j.wear.2008.01.018

Palanikumar, K., and J.P. Davim. 2013. Electrical discharge machining: study on machining characteristics of WC/Co composites. In: J.P. Davim (Ed.), *Machining and Machine-Tools*, pp. 135–168. Philadelphia, PA: Woodhead Publishing. https://doi.org/10.1533/9780857092199.135

Pathak, S., and G.C. Saha. 2017. Development of sustainable cold spray coatings and 3D additive manufacturing components for repair/manufacturing applications: A critical review. *Coatings* 7(8):122. https://doi.org/10.3390/coatings7080122

Pawłowski, L. 2008. *The Science and Engineering of Thermal Spray Coatings* (Second Edition). Chichester, UK: Wiley.

Pecharromán-Gallego, R. 2017. An overview on quantum cascade lasers: Origins and development. In: V.N. Stavrou (Ed.), *Quantum Cascade Lasers.* London: InTechOpen. https://doi.org/10.5772/65003

Pena, C., B. Civit, A. Gallego-Schmid, A. Druckman, A. Caldeira-Pires, B. Weidema, E. Mieras, F. Wang, J. Fava, L.M. Canals, and M. Cordella, 2021. Using life cycle assessment to achieve a circular economy. *The International Journal of Life Cycle Assessment* 1–6. http://dx.doi.org/10.20504/opus2020c2617

Pfähler, K., D. Morar, H.-G. Kemper. 2019. Exploring application fields of additive manufacturing along the product life cycle. *Procedia CIRP* 81:151–156. https://doi.org/10.1016/j.procir.2019.03.027

Phuyal, S., D. Bista, and R. Bista. 2020. Challenges, opportunities and future directions of smart manufacturing: A state of art review. *Sustainable Futures* 2:100023. https://doi.org/10.1016/j.sftr.2020.100023

Pogrebnjak, A., K. Smyrnova, and O. Bondar. 2019. Nanocomposite multilayer binary nitride coatings based on transition and refractory metals: Structure and properties. *Coatings* 9(3):155. https://doi.org/10.3390/coatings9030155

Polli, R., and V. Cook. 1969. Validity of the product life cycle. *The Journal of Business* 42(4):385–400.

Powell, B.R., A.A. Luo, and P.E. Krajewski. 2012. Magnesium alloys for lightweight powertrains and automotive bodies. In: J. Rowe (Ed.), *Advanced Materials in Automotive Engineering*, pp. 150–209. Cambridge, UK: Woodhead Publishing. https://doi.org/10.1533/9780857095466.150

Powell, R.C. 2003. Solid state lasers. In: C. Webb, and J. Jones (Eds.), *Handbook of Laser Technology and Applications*. Vol. 2, pp. 305–306. Boca Raton, FL: CRC Press. https://doi.org/10.1201/NOE0750309608

Pragana, J.P.M., R.F.V. Sampaio, I.M.F. Bragança, C.M.A. Silva, and P.A.F. Martins. 2021. Hybrid metal additive manufacturing: A state–of–the-art review. *Advances in Industrial and Manufacturing Engineering* 2:100032. https://doi.org/10.1016/j.aime.2021.100032

Prasath Manikanda, K., and S. Vignesh. 2017. A review of advanced casting techniques. *Research Journal of Engineering and Technology* 8(4):440–446. https://doi.org/10.5958/2321-581X.2017.00076.9

Prashar, G., H. Vasudev, and L. Thakur. 2020. Performance of different coating materials against slurry erosion failure in hydrodynamic turbines: A review. *Engineering Failure Analysis* 115:104622. https://doi.org/10.1016/j.engfailanal.2020.104622

Prida, V.M., V. Vega, J. García, L. González, W.O. Rosa, A. Fernández, and B. Hernando. 2012. Functional Fe–Pd nanomaterials synthesized by template-assisted methods. *Journal of Magnetism and Magnetic Materials* 324(21):3508–3511. https://doi.org/10.1016/j.jmmm.2012.02.077

Prstić, A., Z. Aćimović-Pavlović, A. Terzić, and L. Pavlović. 2014. Synthesis and characterization of new refractory coatings based on talc, cordierite, zircon and mullite fillers for lost foam casting process. *Archives of Metallurgy and Materials* 59(1):89–95. https://doi.org/10.2478/amm-2014-0015

Pupan, L.I., and V.I. Kononenko. 2008. *Perspectives of Materials Production and Processing*. Kharkov: NTU KhPI (in Russian).

Qian, Y.J., S.W. Han, and H.J. Kwon. 2016. Development of ultrasonic surface treatment device. *Applied Mechanics and Materials* 835:620–625. https://doi.org/10.4028/www.scientific.net/amm.835.620

Qu, M., T. Jin, G. Xie, and R. Cai. 2020. Developing a novel binderless diamond grinding wheel with femtosecond laser ablation and evaluating its performance in grinding soft and brittle materials. *Journal of Materials Processing Technology* 275:116359. https://doi.org/10.1016/j.jmatprotec.2019.116359

Quazi, M.M., M. Ishak, M.A. Fazal, A. Arslan, S. Rubaiee, A. Qaban, M.H. Aiman, T. Sultan, M.M. Ali, and S.M. Manladan. 2020. Current research and development status of dissimilar materials laser welding of titanium and its alloys. *Optics & Laser Technology* 126:106090. https://doi.org/10.1016/j.optlastec.2020.106090

Qudeiri, J.E.A., A. Zaiout, A.-H.I. Mourad, M.H. Abidi, and A. Elkaseer. 2020. Principles and characteristics of different EDM processes in machining tool and die steels. *Applied Sciences* 10:2082. https://doi.org/10.3390/app10062082

Quintino, L. 2014. Overview of coating technologies. In: R. Miranda (Ed.), *Surface Modification by Solid State Processing*, pp. 1–24. Cambridge, UK: Woodhead Publishing. https://doi.org/10.1533/9780857094698.1

Quintino, L., A. Costa, R. Miranda, D. Yapp, V. Kumar, and C.J. Kong. 2007. Welding with high power fiber lasers: A preliminary study. *Materials & Design* 28(4):1231–1237. https://doi.org/10.1016/j.matdes.2006.01.009

Rahaman, M.N., W. Xiao, and W. Huang. 2018. Bioactive glass composites for bone and musculoskeletal tissue engineering. In: H. Ylänen (Ed.), *Bioactive Glasses* (Second Edition), pp. 285–336. Duxford, UK: Woodhead Publishing. https://doi.org/10.1016/B978-0-08-100936-9.00013-7

Rahman, M., Y.S. Wong, and M.D. Nguyen. 2014. Compound and hybrid micromachining: Part II: Hybrid micro-EDM and micro-ECM. *Reference Module in Materials Science and Materials Engineering: Comprehensive Materials Processing* 11:113–150. https://doi.org/10.1016/B978-0-08-096532-1.01329-7

Rao, X., F. Zhang, C. Li, and Y. Li. 2017. Experimental investigation on electrical discharge diamond grinding of RB-SiC ceramics. *The International Journal of Advanced Manufacturing Technology* 94(5–8):2751–2762. 10.1007/s00170-017-1102-7

Rasouli, S., N. Rezaei, H. Hamedi, S. Zendehboudi, and X. Duan. 2021. Superhydrophobic and superoleophilic membranes for oil-water separation application: A comprehensive review. *Materials & Design* 204:109599. https://doi.org/10.1016/j.matdes.2021.109599

Razeghi, M. 2010. *Technology of Quantum Devices*. Boston, MA: Springer. https://doi.org/10.1007/978-1-4419-1056-1_6

Remanufacture. 2021. https://www.remanufacturing.eu/about-remanufacturing.php (accessed on March, 26, 2021).

Ren, N., K. Xia, H. Yang, F. Gao, and Sh. Song. 2021. Water-assisted femtosecond laser drilling of alumina ceramics. *Ceramics International* 47(8):11465–11473. https://doi.org/10.1016/j.ceramint.2020.12.274

Ren, X., L. Jia, and K. Zhang. 2019. Stress–strain relation during superplastic uniaxial tension at constant tension velocity. *Engineering Research Express* 1(1):015035. https://doi.org/10.1088/2631-8695/ab4008

Ribeiro, G.R., I.M.F. Bragança, P.A.R. Rosa, and P.A.F. Martins. 2013. A laboratory machine for micro electrochemical machining. In: J.P. Davim (Ed.), *Machining and Machine-Tools*, pp. 195–210. Cambridge, UK: Woodhead Publishing. https://doi.org/10.1533/9780857092199.195

Richter, S., F. Zimmermann, A. Tünnermann, and S. Nolte. 2016. Laser welding of glasses at high repetition rates: Fundamentals and prospects. *Optics & Laser Technology* 83:59–66. https://doi.org/10.1016/j.optlastec.2016.03.022

Rico, A., A. Salazar, M.E. Escobar, J. Rodriguez, and P. Poza. 2018. Optimization of atmospheric low-power plasma spraying process parameters of Al_2O_3-50wt%Cr_2O_3 coatings. *Surface and Coatings Technology* 354:281–296. https://doi.org/10.1016/j.surfcoat.2018.09.032

Rohleder, M., and F. Jakob. 2016. Foam injection molding. In: H.-P. Heim (Ed.), *Specialized Injection Molding Techniques*, pp. 53–106. Oxford, UK: William Andrew Publishing. https://doi.org/10.1016/B978-0-323-34100-4.00002-X

Romanov, P.P. 2010. Technology of vacuum deposition. *The World of Technic and Technology* 3(100):18–19 (in Russian).

Rosato, D.V., D.V. Rosato, and M.V. Rosato. 2004. *Plastic Product Material and Process Selection Handbook*. Oxford, UK: Elsevier. https://doi.org/10.1016/B978-1-85617-431-2.X5000-2

Rossi, S., F. Russo, and M. Calovi. 2021. Durability of vitreous enamel coatings and their resistance to abrasion, chemicals, and corrosion: A review. *Journal of Coatings Technology and Research* 18:39–52. https://doi.org/10.1007/s11998-020-00415-3

Roy, J., S. Chandra, S. Das, and S. Maitra. 2014. Oxidation behaviour of silicon carbide: A review. *Reviews on Advanced Materials Science* 38:29–39.

Rusakov, P.V., and O.I. Shinsky. 2014. Method of vibration and gasodynamic compaction of the mold with identification of the sand densification. *Foundry Processes* 103(1):55–61 (in Russian).

Sabard, A., and T. Hussain. 2019. Inter-particle bonding in cold spray deposition of a gas-atomised and a solution heat-treated Al 6061 powder. *Journal of Materials Science* 54:12061–12078. https://doi.org/10.1007/s10853-019-03736-w

Sadikov, I.R., and V.D. Mogilevets. 2016. Application of laser tehnologies for cleaning, cladding, and heat treatment of the forgings using the 3rd generation robot techonolgies. In: Proceedings of Engineering Techniques and Technology Conference, 04–10 April 2016, Omsk, Russia, pp. 304–310.

Sahoo, A., and S. Tripathy. 2021. Development in plasma arc welding process: A review. *Materials Today: Proceedings* 41(Part 2):363–368. https://doi.org/10.1016/j.matpr.2020.09.562

Samal, S. 2017. Thermal plasma technology: The prospective future in material processing. *Journal of Cleaner Production* 142(Part 4):3131–3150. https://doi.org/10.1016/j.jclepro.2016.10.154

Sanchez, F.A.C., H. Boudaoud, M. Camargo, and J.M. Pearce. 2020. Plastic recycling in additive manufacturing: A systematic literature review and opportunities for the circular economy. *Journal of Cleaner Production* 264:121602. https://doi.org/10.1016/j.jclepro.2020.121602

Santecchia, E., A.M.S. Hamouda, F. Musharavati, E. Zalnezhad, M. Cabibbo, and S. Spigarelli. 2015. Wear resistance investigation of titanium nitride-based coatings. *Ceramics International* 41(9, Part A):10349–10379. https://doi.org/10.1016/j.ceramint.2015.04.152

Saravanan, S., V. Vijayan, S.T. Jaya Suthahar, A.V. Balan, S. Sankar, and M. Ravichandran. 2020. A review on recent progresses in machining methods based on abrasive water jet machining. *Materials Today: Proceedings* 21(Part 1):116–122. https://doi.org/10.1016/j.matpr.2019.05.373

Sarri, G. 2015. World's most powerful laser is 2,000 trillion watts: But what's it for? https://theconversation.com/worlds-most-powerful-laser-is-2-000-trillion-watts-but-whats-it-for-45891 (accessed on February, 25, 2021).

Sawant, T.S.S. 2020. Ultrasonic machining process. *International Research Journal of Engineering and Technology* 7(4):140–141.

Saxena, K.K., M. Bellotti, D. Van Camp, J. Qian, and D. Reynaerts. 2018. Electrochemical based hybrid machining. In: X. Luo, and Y. Qin (Eds.), *Hybrid Machining: Theory, Methods, and Case Studies*, pp. 111–129. London: Academic Press. https://doi.org/10.1016/B978-0-12-813059-9.00005-1

Schaefer, D. 2018. Basics of ion beam figuring and challenges for real optics treatment. *Proc. SPIE* 10829:1082907. https://doi.org/10.1117/12.2318572

Schäfer, G. 2010. Degradation of glass linings and coatings. In: B. Cottis, M. Graham, R. Lindsay, S. Lyon, T. Richardson, D. Scantlebury, H. Stott (Eds.), *Shreir's Corrosion*, pp. 2319–2329. Amsterdam: Elsevier. https://doi.org/10.1016/B978-044452787-5.00178-5

Schmüser, P. 2003. Free-electron lasers. In: CERN Accelerator School and DESY Zeuthen: Accelerator Physics, 15-26 September 2003, Zeuthen, Germany, pp. 477–493.

Schramm, A., F. Morczinek, U. Götze, and M. Putz. 2020. Technical-economic evaluation of abrasive recycling in the suspension fine jet process chain. *The International Journal of Advanced Manufacturing Technology* 106:981–992. https://doi.org/10.1007/s00170-019-04651-9

Schulze, H.P. 2017. Importance of polarity change in the electrical discharge machining. *AIP Conference Proceedings* 1896(1):050001. https://doi.org/10.1063/1.5008046

Schumacher, B.M., R. Krampitz, and J.-P. Kruth. 2013. Historical phases of EDM development driven by the dual influence of "Market Pull" and "Science Push." *Procedia CIRP* 6:5–12. doi:10.1016/j.procir.2013.03.001

Schwartz, M. 2010. *New Materials, Processes, and Methods Technology.* Boca Raton, FL: CRC Press. https://doi.org/10.1201/9781420039344

Sebastian, A., F. Zhang, A. Dodda, D. May-Rawding, H. Liu, T. Zhang, M. Terrones, and S. Das. 2019. Electrochemical polishing of two-dimensional materials. *ACS Nano* 13(1):78–86. https://doi.org/10.1021/acsnano.8b08216

Segal, V. 2018. Review: Modes and processes of severe plastic deformation (SPD). *Materials* 11:1175. https://doi.org/10.3390/ma11071175
Semenikhin, B.A., L.P. Kuznetsova, and V.I. Kozlikin. 2020. The use of hard alloy waste in composite galvanic coatings for the restoration of car parts. *Solid State Phenomena* 299:258–263. https://doi.org/10.4028/www.scientific.net/ssp.299.258
Senthil, J., M. Prabhahar, C. Thiagarajan, S. Prakash, and R. Lakshmanan. 2020. Studies on performance and process improvement of implementing novel vacuum process for new age castings. *Materials Today: Proceedings* 33(Part 1):813–819. https://doi.org/10.1016/j.matpr.2020.06.269
Seo, D., M. Sayar, and K. Ogawa. 2012. SiO_2 and $MoSi_2$ formation on Inconel 625 surface via SiC coating deposited by cold spray. *Surface and Coatings Technology* 206(11–12):2851–2858. https://doi.org/10.1016/j.surfcoat.2011.12.010
Shang, S.M., and W. Zeng. 2013. Conductive nanofibres and nanocoatings for smart textiles. In: T. Kirstein (Ed.), *Multidisciplinary Know-How for Smart-Textiles Developers*, pp. 92–128. Cambridge, UK: Woodhead Publishing. https://doi.org/10.1533/9780857093530.1.92
Shekhar, M., and S.K.S. Yadav. 2020. Diamond abrasive based cutting tool for processing of advanced engineering materials: A review. *Materials Today: Proceedings* 22 (Part 4):3126–3135. https://doi.org/10.1016/j.matpr.2020.03.449
Shen, L., J. Zhao, Y. Zhang, and G. Quan. 2021. Performance evaluation of titanium-based metal nitride coatings and die lifetime prediction in a cold extrusion process. *High Temperature Materials and Processes* 40(1):108–120. https://doi.org/10.1515/htmp-2021-0019
Shoemaker, R. 2003. Principles. In: C. Webb, and J. Jones (Eds.), *Handbook of Laser Technology and Applications*. Vol. 1, pp. 3–4. Boca Raton, FL: CRC Press. https://doi.org/10.1201/NOE0750309608
Shourgeshty, M., M. Aliofkhazraei, and M.M. Alipour. 2016. *Introduction to High-Temperature Coatings*. London: InTechOpen. https://doi.org/10.5772/64282
Shri Prakash, B., S. Senthil Kumar, and S.T. Aruna. 2017. Microstructure and performance of LSM/YSZ based solid oxide fuel cell cathodes fabricated from solution combustion co-synthesized powders and by solution precursor plasma spraying. *Surface and Coatings Technology* 310:25–32. https://doi.org/10.1016/j.surfcoat.2016.12.004
Shrivastava, A., S.A. Kumar, R. Samrat, B.K. Nagesha, B. Sanjay, and T.N. Suresh. 2021. Remanufacturing of nickel-based aero-engine components using metal additive manufacturing technology. *Materials Today: Proceedings* 45(Part 6):4893–4897. https://doi.org/10.1016/j.matpr.2021.01.355
Shukla, P., and J. Lawrence. 2015. Characterization and modification of technical ceramics through laser surface engineering. In: J. Lawrence, and D.G. Waugh (Eds.), *Laser Surface Engineering*, pp. 107–134. Cambridge, UK: Woodhead Publishing. https://doi.org/10.1016/B978-1-78242-074-3.00005-2
Shutov, F.A., G. Henrici-Olivé, and S. Olivé. 1986. Injection molding: Gas counter pressure process. In: Henrici-Olivé G., and S. Olivé (Eds.), *Integral/Structural Polymer Foams*. Berlin – Heidelberg: Springer. https://doi.org/10.1007/978-3-662-02486-7_6
Siddiqui, A.A., and A.K. Dubey. 2021. Recent trends in laser cladding and surface alloying. *Optics & Laser Technology* 134:106619. https://doi.org/10.1016/j.optlastec.2020.106619
Sikarwar, V.S., M. Hrabovský, G. Van Oost, M. Pohořelý, and M. Jeremiáš. 2020. Progress in waste utilization via thermal plasma. *Progress in Energy and Combustion Science* 81:100873. https://doi.org/10.1016/j.pecs.2020.100873
Simonenko, E.P., D.V. Sevast'yanov, N.P. Simonenko, V.G. Sevast'yanov, and N.T. Kuznetsov. 2013. Promising ultra-high-temperature ceramic materials for aerospace applications. *Russian Journal of Inorganic Chemistry* 58:1669–1693. https://doi.org/10.1134/S0036023613140039

Singh, H., and P.K. Jain. 2016. Remanufacturing of functional surfaces using developed ECH machine. *Journal of Remanufacturing* 6:2. https://doi.org/10.1186/s13243-016-0024-0

Singh, K.J., I.S. Ahuja, and J. Kapoor. 2017. Chemical assisted ultrasonic machining of polycarbonate glass and optimization of process parameters by Taguchi and Grey relational analysis. *Advances in Materials and Processing Technologies* 3(4):563–585. DOI: 10.1080/2374068X.2017.1350022

Singh, R., A. Gupta, O. Tripathi, S. Srivastava, Bh. Singh, A. Awasthi, S.K. Rajput, P. Sonia, P. Singhal, and K.K. Saxena. 2020. Powder bed fusion process in additive manufacturing: An overview. *Materials Today: Proceedings* 26(Part 2)3058–3070. https://doi.org/10.1016/j.matpr.2020.02.635

Singh, R.P., and S. Singhal. 2016. Rotary ultrasonic machining: A review. *Materials and Manufacturing Processes* 31(14):1795–1824. https://doi.org/10.1080/10426914.2016.1140188

Smirnov, V.M., E.P. Shalunov, and I.S. Golyushov. 2020. Physical and mechanical properties and structure of copper- based composite materials for diamond tools binder. *Journal of Physics: Conference Series* 1431:012054. https://doi.org/10.1088/1742-6596/1431/1/012054

Sobiesierski, A., and P.M. Smowton. 2016. Quantum-dot lasers: Physics and applications. *Reference Module in Materials Science and Materials Engineering*. Amsterdam: Elsevier. https://doi.org/10.1016/B978-0-12-803581-8.00836-5,

Sokołowski, P., P. Nylen, R. Musalek, L. Łatka, S. Kozerski, D. Dietrich, T. Lampke, and L. Pawłowski. 2017. The microstructural studies of suspension plasma sprayed zirconia coatings with the use of high-energy plasma torches. *Surface and Coatings Technology* 318:250–261. https://doi.org/10.1016/j.surfcoat.2017.03.025

Song, W., Zh. Li, Y. Li, H. You, P. Qi, F. Liu, and D.A. Loy. 2018. Facile sol-gel coating process for anti-biofouling modification of poly (vinylidene fluoride) microfiltration membrane based on novel zwitterionic organosilica. *Journal of Membrane Science* 550:266–277. https://doi.org/10.1016/j.memsci.2017.12.076

Spaeth, M.L., K.R. Manes, D.H. Kalantar, P.E. Miller, J.E. Heebner, E.S. Bliss, D.R. Spec, T.G. Parham, P.K. Whitman, P.J. Wegner, et al. 2016. Description of the NIF Laser. *Fusion Science and Technology* 69(1):25–145. https://doi.org/10.13182/FST15-144

Speidel, A., R. Sélo, I. Bisterov, J. Mitchell-Smith, and A.T. Clare. 2021. Post processing of additively manufactured parts using electrochemical jet machining. *Materials Letters* 292:129671. https://doi.org/10.1016/j.matlet.2021.129671

Sreejith, P., and B. Ngoi. 2001. New materials and their machining. *The International Journal of Advanced Manufacturing Technology* 18:537–544. https://doi.org/10.1007/s001700170030

Srivastava, A.K., N. Kumar, and A.R. Dixit. 2021. Friction stir additive manufacturing – An innovative tool to enhance mechanical and microstructural properties. *Materials Science and Engineering: B* 263:114832. https://doi.org/10.1016/j.mseb.2020.114832

Staniewicz-Brudnik, B., E. Bączek, and G. Skrabalak. 2015. The new generation of diamond wheels with vitrified (ceramic) bonds. In: A. Lakshmanan (Ed.), *Sintering Techniques of Materials*. Rijeka: IntechOpen. https://doi.org/10.5772/59503

Steen, W.M. 2003. *Laser Material Processing* (3rd Edition). London: Springer-Verlag.

Stepanov, A.L. 2019. Optical properties of polymer nanocomposites with functionalized nanoparticles. In: K. Pielichowski, and T.M. Majka (Eds.), *Polymer Composites with Functionalized Nanoparticles*, pp. 325–355. Amsterdam: Elsevier. https://doi.org/10.1016/B978-0-12-814064-2.00010-X

Stevens, J.E. 2000. Plasma fundamentals for materials processing. In: R.J. Shul, and S.J. Pearton (Eds.), *Handbook of Advanced Plasma Processing Techniques*, pp. 33–68. Berlin – Heidelberg: Springer-Verlag.

Su, H.H., H.J. Xu, B. Xiao, Y.C. Fu, and J.H. Xu. 2006. Microstructure and performance of porous Ni-Cr alloy bonded diamond grinding wheel. *Materials Science Forum* 532–533:373–376. https://doi.org/10.4028/www.scientific.net/msf.532-533.373

Šulc, J., and H. Jelínková. 2013. Solid-state lasers for medical applications. In: H. Jelínková (Ed.), *Lasers for Medical Applications*, pp. 127–176. Cambridge, UK: Woodhead Publishing. https://doi.org/10.1533/9780857097545.2.127

Sun, H., and K. Kusumoto. 2010. Drilling performance of plasma arc drilling method. *Quarterly Journal of the Japan Welding Society* 28(4):421–426. https://doi.org/10.2207/qjjws.28.421

Sun, Y., Ch. Zhang, J. Wu, Q. Meng, B. Liu, K. Gao, and L. He. 2019. Enhancement of oxidation resistance via titanium boron carbide coatings on diamond particles. *Diamond and Related Materials* 92:74–80. https://doi.org/10.1016/j.diamond.2018.12.019

Sykutera, D., P. Czyżewski, and P. Szewczykowski. 2020. The microcellular structure of injection molded thick-walled parts as observed by in-line monitoring. *Materials* 13:5464. https://doi.org/10.3390/ma13235464

Sze, R.C., and D.G. Harris. 1995. Tunable excimer lasers. In: F.J. Duarte (Ed.), *Tunable Lasers Handbook*, pp. 33–61. San Diego: Academic Press. https://doi.org/10.1016/B978-012222695-3/50004-0

Szostak, M., P. Krzywdzińska, and M. Barczewski. 2018. MuCell® and InduMold technologies in production of high quality automotive parts from polymer materials. *Polimery* 63(2):145–152. https://doi.org/dx.doi.org/10.14314/polimery.2018.2.8

Teixeira, V., J. Carneiro, P. Carvalho, E. Silva, S. Azevedo, and C. Batista. 2011. High barrier plastics using nanoscale inorganic films. In: J.-M. Lagarón (Ed.), *Multifunctional and Nanoreinforced Polymers for Food Packaging*, pp. 285–315. Cambridge, UK: Woodhead Publishing. https://doi.org/10.1533/9780857092786.1.285

The Library of Manufacturing. 2021. http://thelibraryofmanufacturing.com/vacuum_mold_casting.html (accessed on February, 8, 2021).

Thoe, T.B., D.K. Aspinwall, and M.L.H. Wise. 1998. Review on ultrasonic machining. *International Journal of Machine Tools and Manufacture* 38(4):239–255. https://doi.org/10.1016/S0890-6955(97)00036-9

TiN Coatings (Titanium Nitride). 2021. http://www.teercoatings.co.uk/index.php?page=9 (accessed on June, 14, 2021)

Tolio, T., A. Bernard, M. Colledani, S. Kara, G. Seliger, J. Duflou, O. Battaia, and Sh. Takata. 2017. Design, management and control of demanufacturing and remanufacturing systems. *CIRP Annals* 66(2):585–609. https://doi.org/10.1016/j.cirp.2017.05.001

Tominaga, K., H. Adachi, and K. Wasa. 2012. Functional thin films. In: K. Wasa, I. Kanno, and H. Kotera (Eds.), *Handbook of Sputtering Technology* (Second Edition), pp. 361–520. Oxford, UK: William Andrew Publishing. https://doi.org/10.1016/B978-1-4377-3483-6.00006-1

Tomlinson, D., and J. Wichmann. 2014. Chemical milling environmental improvements, aerospace is green and growing. *NASF Surface Technology White Papers* 78(9):1–7.

Toumanov, N.I. 2003. *Plasma and High Frequency Processes for Obtaining and Processing Materials in the Nuclear Fuel Cycle*. New York: Nova Science Publishers.

Trexel Data on MuCell Technology. 2021. https://trexel.com/technology-solutions/mucell/ (accessed on February, 4, 2021).

Tripathi, H.S., B. Mukherjee, S. Das, M.K. Haldar, S.K. Das, and A. Ghosh. 2003. Synthesis and densification of magnesium aluminate spinel: Effect of MgO reactivity. *Ceramics International* 29(8):915–918. https://doi.org/10.1016/S0272-8842(03)00036-1

Tsegelnik, E.V. 2015. Perspectives of the laser technology applications in the aircraft industry. *Open Infromation and Computer Integrated Technologies* 70:121–129 (in Russian).

Tu, X., Sh. Xie, Y. Zhou, and L. He. 2021. Surface gradient heterogeneity induced tensile plasticity in a Zr-based bulk metallic glass through ultrasonic impact treatment. *Journal of Non-Crystalline Solids* 554:120612. https://doi.org/10.1016/j.jnoncrysol.2020.120612

Tyczynski, P., Z. Siemiatkowski, and M. Rucki. 2019. Drill base body fabricated with additive manufacturing technology: structure, strenght and reliability. *Journal of Machine Construction and Maintenance* 113(2):59–65.

Uhlmann, E., and C. Männel. 2019. Modelling of abrasive water jet cutting with controlled depth for near-net-shape fabrication. *Procedia CIRP* 81:920–925. https://doi.org/10.1016/j.procir.2019.03.228

Uhlmann, E., M. Röhner, and M. Langmack. 2010. Micro-EDM/. In: Y. Qin (Ed.), *Micro-Manufacturing Engineering and Technology*, pp. 39–58. Oxford, UK: William Andrew Publishing. https://doi.org/10.1016/B978-0-8155-1545-6.00003-X

Unune, D.R., Ch.K. Nirala, and H.S. Mali. 2018. ANN-NSGA-II dual approach for modeling and optimization in abrasive mixed electro discharge diamond grinding of Monel K-500. *Engineering Science and Technology* 21(3):322–329. https://doi.org/10.1016/j.jestch.2018.04.014

Uzunyan, M.D. 2003. *Diamond Spark Grinding of Hard Alloys*. Kharkov: NTU "KhPI" (in Russian).

Vaidya, S., P. Ambad, and S. Bhosle. 2018. Industry 4.0: A Glimpse. *Procedia Manufacturing* 20:233–238. https://doi.org/10.1016/j.promfg.2018.02.034

van Hove, R.P., I.N. Sierevelt, B.J. van Royen, and P.A. Nolte. 2015. Titanium-nitride coating of orthopaedic implants: A review of the literature. *BioMed Research International* 2015:485975. https://doi.org/10.1155/2015/485975

Vandenabeele, C.R., and S. Lucas. 2020. Technological challenges and progress in nanomaterials plasma surface modification: A review. *Materials Science and Engineering: R: Reports* 139:100521. https://doi.org/10.1016/j.mser.2019.100521

Vasil'ev, E.V., and A.Y. Popov. 2015. Restoration of an axial hard-alloy tool by deep diamond grinding. *Russian Engineering Research* 35:780–782. https://doi.org/10.3103/S1068798X15100251

Vaßen, R., K.-H. Rauwald, O. Guillon, J. Aktaa, T. Weber, H.C. Back, D. Qu, and J. Gibmeier. 2018. Vacuum plasma spraying of functionally graded tungsten/EUROFER97 coatings for fusion applications. *Fusion Engineering and Design* 133:148–156. https://doi.org/10.1016/j.fusengdes.2018.06.006

Velayudham, A. 2007. Modern manufacturing processes: A review. *Journal on Design and Manufacturing Technologies* 1(1):30–40.

Verlinden, B., D. Cattrysse, and D. Van Oudheusden. 2007. Integrated sheet-metal production planning for laser cutting and bending. *International Journal of Production Research* 45(2):369–383. https://doi.org/10.1080/00207540600658062

Villamil Jiménez, J.A., N. Le Moigne, J.-C. Bénézet, M. Sauceau, R. Sescousse, and J. Fages. 2020. Foaming of PLA composites by supercritical fluid-assisted processes: A review. *Molecules* 25(15):3408. https://doi.org/10.3390/molecules25153408

Vincent, R., M. Langlotz, and M. Düngen. 2020. Viscosity measurement of polypropylene loaded with blowing agents (propane and carbon dioxide) by a novel inline method. Journal of Cellular Plastics 56(1):73–88. https://doi.org/10.1177/0021955X19864400

von Niessen, K., and M. Gindrat. 2011. Plasma spray-PVD: A new thermal spray process to deposit out of the vapor phase. *Journal of Thermal Science and Technology* 20:736–743. https://doi.org/10.1007/s11666-011-9654-9

Vovk, R.V., E.S. Gevorkyan, V.P. Nerubatskyi, M.M. Prokopiv, V.O. Chishkala, and O.M. Melnik. 2018. *New Instrumental Ceramic Composites*. Kharkiv: V. N. Karazin Kharkiv National University (in Ukrainian).

Wahab, D.A., E. Blanco-Davis, A.K. Ariffin, and J. Wang. 2018. A review on the applicability of remanufacturing in extending the life cycle of marine or offshore components and structures. *Ocean Engineering* 169:125–133. https://doi.org/10.1016/j.oceaneng.2018.08.046

Wallace, J. 2016. High-power fiber lasers: Kilowatt-level fiber lasers mature. *Laser Focus World.* https://static1.squarespace.com/static/5d5d8be5c16a590001b58605/t/5e414b7b2ed8a6567279527f/1581337469234/high-power-fiber-lasers-techniques-and-accessories_whitepaperpdf_render.pdf (accessed on February, 25, 2021).

Wang, B., F. Tao, X. Fang, Ch. Liu, Y. Liu, and Th. Freiheit. 2021. Smart manufacturing and intelligent manufacturing: A comparative review. *Engineering*, Article in Press. https://doi.org/10.1016/j.eng.2020.07.017

Wang, C., Y. Wang, S. Fan, Y. You, L. Wang, C. Yang, X. Sun, and X. Li. 2015. Optimized functionally graded La2Zr2O7/8YSZ thermal barrier coatings fabricated by suspension plasma spraying. *Journal of Alloys and Compounds* 649:1182–1190. https://doi.org/10.1016/j.jallcom.2015.05.290

Wang, H., Y. Xu, H. Zheng, W. Zhou, N. Ren, X. Ren, and T. Li. 2019. Monitoring and analysis of millisecond laser drilling process and performance with and without longitudinal magnetic assistance and/or assist gas. *Journal of Manufacturing Processes* 48:297–312. https://doi.org/10.1016/j.jmapro.2019.10.015

Wang, J., J. Yu, Y. Yang, F. Du, and J. Wang. 2020. Study of the remanufacturing critical threshold and remanufacturability evaluation for FV520B-I blade based on fatigue life and FEA. *Engineering Failure Analysis* 112:104509. https://doi.org/10.1016/j.engfailanal.2020.104509

Wang, K., Ch.-Ch. Ho, Ch. Zhang, and B. Wang. 2017. A review on the 3D printing of functional structures for medical phantoms and regenerated tissue and organ applications. *Engineering* 3(5):653–662. https://doi.org/10.1016/J.ENG.2017.05.013

Wang, X., L. Jiang, D. Zhang, T.J. Rupert, I.J. Beyerlein, S. Mahajan, E.J. Lavernia, and J.M. Schoenung. 2020. Revealing the deformation mechanisms for room-temperature compressive superplasticity in nanocrystalline magnesium. *Materialia* 11:100731. https://doi.org/10.1016/j.mtla.2020.100731\

Wang, Y., K. Li, Sh. Gan, and Ch. Cameron. 2019. Analysis of energy saving potentials in intelligent manufacturing: A case study of bakery plants. *Energy* 172:477–486. https://doi.org/10.1016/j.energy.2019.01.044

Webb, C., and J. Jones (Eds.). 2003. *Handbook of Laser Technology and Applications.* Boca Raton, FL: CRC Press. https://doi.org/10.1201/NOE0750309608

Węglowski, M.S. 2018. Friction stir processing: State of the art. *Archives of Civil and Mechanical Engineering* 18(1):114–129. https://doi.org/10.1016/j.acme.2017.06.002

Węglowski, M.S., S. Błacha, and A. Phillips. 2016. Electron beam welding: Techniques and trends: Review. *Vacuum* 130:72–92. https://doi.org/10.1016/j.vacuum.2016.05.004

Wegmann, H., M. Holthaus, F.J. Guesthuisen. 2010. Plasma cutting as an effective processing process. *The World of Technic and Technology* 10(107):26–29 (in Russian).

Weingärtner, E., F. Kuster, and K. Wegener. 2012. Modeling and simulation of electrical discharge machining. *Procedia CIRP* 2:74–78. https://doi.org/10.1016/j.procir.2012.05.043

Werner, A., N. Zimmermann, and J. Lentes. 2019. Approach for a holistic predictive maintenance strategy by incorporating a digital twin. *Procedia Manufacturing* 39:1743–1751. https://doi.org/10.1016/j.promfg.2020.01.265

Wilde, G. 2014. Physical metallurgy of nanocrystalline metals. In: D.E. Laughlin, K. Hono (Eds.), *Physical Metallurgy* (Fifth Edition), pp. 2707–2805. Amsterdam: Elsevier. https://doi.org/10.1016/B978-0-444-53770-6.00026-5

Willemin, S., P. Carminati, S. Jacques, J. Roger, and F. Rebillat. 2017. Identification of complex oxidation/corrosion behaviours of boron nitride under high temperature. *Oxidation of Metals* 88(3–4):247–256. https://doi.org/10.1007/s11085-017-9739-z

Wong, A., H. Guo, V. Kumar, Ch.B. Park, and N.P. Suh. 2016. Microcellular plastics. In: H.F. Mark (Ed.), *Encyclopedia of Polymer Science and Technology*. Hoboken, NJ: Wiley. https://doi.org/10.1002/0471440264.pst468.pub2

Wu, C.S., L. Wang, W.J. Ren, and X.Y. Zhang. 2014. Plasma arc welding: Process, sensing, control and modeling. *Journal of Manufacturing Processes* 16(1):74–85. https://doi.org/10.1016/j.jmapro.2013.06.004

Wu, W., J. Lu, and H. Zhang. 2019. Smart factory reference architecture based on CPS fractal. *IFAC-PapersOnLine* 52(13):2776–2781. https://doi.org/10.1016/j.ifacol.2019.11.628

Wu, Z., T. Li, Q. Li, B. Shi, X. Li, X. Wang, H. Lu, and H. Zhang. 2019. Process optimization of laser cladding Ni60A alloy coating in remanufacturing. *Optics & Laser Technology* 120:105718. https://doi.org/10.1016/j.optlastec.2019.105718

Wypych, G. 2017. *Handbook of Foaming and Blowing Agents*. Toronto: ChemTec Publishing. https://doi.org/10.1016/B978-1-895198-99-7.50013-1

Xia, K., H. Wang, N. Ren, X. Ren, D. Liu, Ch. Shi, T. Li, and J. Tian. 2021. Laser drilling in nickel super-alloy sheets with and without ultrasonic assistance characterized by transient in-process detection with indirect characterization after hole-drilling. *Optics & Laser Technology* 134:106559. https://doi.org/10.1016/j.optlastec.2020.106559

Xiao, L., W. Liu, Q. Guo, L. Gao, G. Zhang, and X. Chen. 2018. Comparative life cycle assessment of manufactured and remanufactured loading machines in China. *Resources, Conservation and Recycling* 131:225–234. https://doi.org/10.1016/j.resconrec.2017.12.021

Xiao, R., and X. Zhang. 2014. Problems and issues in laser beam welding of aluminum–lithium alloys. *Journal of Manufacturing Processes* 16(2):166–175. https://doi.org/10.1016/j.jmapro.2013.10.005

Xie, M., Ch. Zhu, and J. Zhou. 2015. Mold-filling and solidification simulation of grey iron in lost-foam casting. In: Proceedings of the 5th International Conference on Advanced Design and Manufacturing Engineering ICADME 2015, pp. 387–394. https://doi.org/10.2991/icadme-15.2015.78

Xu, J. 2010. *Microcellular Injection Molding*. Hoboken, NJ: Wiley.

Xu, P., G. Meng, L. Pershin, J. Mostaghimi, and Th.W. Coyle. 2021. Control of the hydrophobicity of rare earth oxide coatings deposited by solution precursor plasma spray by hydrocarbon adsorption. *Journal of Materials Science & Technology* 62:107–118. https://doi.org/10.1016/j.jmst.2020.04.044

Xu Z.W., F. Fang, and G. Zeng. 2015. Focused ion beam nanofabrication technology. In: *Handbook of Manufacturing Engineering and Technology*, pp 1391–1423. London: Springer.

Yan, J., and N. Takayama. 2020. *Micro and Nanoscale Laser Processing of Hard Brittle Materials*. Amsterdam: Elsevier. https://doi.org/10.1016/B978-0-12-816709-0.00002-1

Yang, J., L.-L. Ke, and S. Kitipornchai. 2009. Thermo-mechanical analysis of an inhomogeneous double-layer coating system under hertz pressure and tangential traction. *Mechanics of Advanced Materials and Structures* 16(4):308–318. https://doi.org/10.1080/15376490802666302

Yang, M., H. Liu, B. Ye, and W. Qian. 2021. Recycling of printed circuit boards by abrasive waterjet cutting. *Process Safety and Environmental Protection* 148:805–812. https://doi.org/10.1016/j.psep.2021.01.052

Yang, T., W. Ma, X. Meng, W. Huang, Y. Bai, and H. Dong. 2020. Deposition characteristics of CeO_2-Gd2O3 co-stabilized zirconia (CGZ) coating prepared by solution precursor plasma spray. *Surface and Coatings Technology* 381:125114. https://doi.org/10.1016/j.surfcoat.2019.125114

Ye, H., J. Zhu, Y. Liu, W. Liu, and D. Wang. 2020. Microstructure and mechanical properties of laser cladded CrNi alloy by hard turning (HT) and ultrasonic surface rolling (USR). *Surface and Coatings Technology* 393:125806. https://doi.org/10.1016/j.surfcoat.2020.125806

Yip, W.S., S. To, and Zh. Sun. 2021. Hybrid ultrasonic vibration and magnetic field assisted diamond cutting of titanium alloys. *Journal of Manufacturing Processes* 62:743–752. https://doi.org/10.1016/j.jmapro.2020.12.037

Yogeswaran, R., and P. Pitchipoo. 2020. Characterization and machining analysis of AA3003 honeycomb sandwich. *Materials Today: Proceedings* 28(Part 1):4–7. https://doi.org/10.1016/j.matpr.2019.12.101

Yu, Z.X., Y.Z. Ma, Y.L. Zhao, J.B. Huang, W.Z. Wang, M. Moliere, and H.L. Liao. 2017. Effect of precursor solutions on ZnO film via solution precursor plasma spray and corresponding gas sensing performances. *Applied Surface Science* 412:683–689. https://doi.org/10.1016/j.apsusc.2017.03.217

Zenisek, J., N. Wild, and J. Wolfartsberger. 2021. Investigating the potential of smart manufacturing technologies. *Procedia Computer Science* 180:507–516. https://doi.org/10.1016/j.procs.2021.01.269

Zeuner, M., and S. Kontke. 2012. Ion beam figuring technology in optics manufacturing. *Optik & Photonik* 2:56–58.

Zhang, H.-C., C.-N. Yu, Y. Liang, G.-X. Lin, and C. Meng. 2019. Foaming behavior and microcellular morphologies of incompatible SAN/CPE blends with supercritical carbon dioxide as a physical blowing agent. *Polymers* 11(1):89. https://doi.org/10.3390/polym11010089

Zhang, J., Y. Jian, X. Zhao, D. Meng, F. Pan, and Q. Han. 2021. The tribological behavior of a surface-nanocrystallized magnesium alloy AZ31 sheet after ultrasonic shot peening treatment. *Journal of Magnesium and Alloys*. Article in press. https://doi.org/10.1016/j.jma.2020.11.012

Zhang, J.Sh. 2010. *High Temperature Deformation and Fracture of Materials*. Cambridge, UK: Woodhead Publishing. https://doi.org/10.1533/9780857090805.1.154

Zhang, P., X. Guo, X. Ren, Zh. Chen, and Ch. Shen. 2018. Development of $Mo(Si,Al)_2$-MoB composite coatings to protect TZM alloy against oxidation at 1400 °C. *Intermetallics* 93:134–140. https://doi.org/10.1016/j.intermet.2017.12.002

Zhang, X., and X. Ming. 2021. An implementation for smart manufacturing information system (SMIS) from an industrial practice survey. *Computers & Industrial Engineering* 151:106938. https://doi.org/10.1016/j.cie.2020.106938

Zhang, X., W. Fan, and T. Liu. 2020. Fused deposition modeling 3D printing of polyamide-based composites and its applications. *Composites Communications* 21:100413. https://doi.org/10.1016/j.coco.2020.100413

Zhang, Y., and X. Xu. 2019. Influence of surface topography evolution of grinding wheel on the optimal material removal rate in grinding process of cemented carbide. *International Journal of Refractory Metals and Hard Materials* 80:130–143. https://doi.org/10.1016/j.ijrmhm.2019.01.009

Zhang, Y.-L., Q.-D. Chen, H. Xia, and H.-B. Sun. 2010. Designable 3D nanofabrication by femtosecond laser direct writing. *Nano Today* 5(5):435–448. https://doi.org/10.1016/j.nantod.2010.08.007

Zhang, Z.P., M.Zh. Rong, and M.Q. Zhang. 2018. Polymer engineering based on reversible covalent chemistry: A promising innovative pathway towards new materials and new functionalities. *Progress in Polymer Science* 80:39–93. https://doi.org/10.1016/j.progpolymsci.2018.03.002

Zhao, J., Y. Qiao, G. Wang, Ch. Wang, and Ch.B. Park. 2020. Lightweight and tough PP/talc composite foam with bimodal nanoporous structure achieved by microcellular injection molding. *Materials & Design* 195:109051. https://doi.org/10.1016/j.matdes.2020.109051

Zhao, J., J. Huang, Y. Xiang, R. Wang, X. Xu, Sh. Ji, and W. Hang. 2021. Effect of a protective coating on the surface integrity of a microchannel produced by microultrasonic machining. *Journal of Manufacturing Processes* 61:280–295. https://doi.org/10.1016/j.jmapro.2020.11.027

Zhao, J., Zh. Liu, B. Wang, J. Hu, and Y. Wan. 2021. Tool coating effects on cutting temperature during metal cutting processes: Comprehensive review and future research directions. *Mechanical Systems and Signal Processing* 150:107302. https://doi.org/10.1016/j.ymssp.2020.107302

Zhao, X. 2011. Bioactive materials in orthopaedics. In: X. Zhao, J.M. Courtney, and H. Qian (Eds.), *Bioactive Materials in Medicine*, pp. 124–154. Cambridge, UK: Woodhead Publishing. https://doi.org/10.1533/9780857092939.2.124

Zheng, H., E. Li, Y. Wang, P. Shi, B. Xu, and Sh. Yang. 2019. Environmental life cycle assessment of remanufactured engines with advanced restoring technologies. *Robotics and Computer-Integrated Manufacturing* 59:213–221. https://doi.org/10.1016/j.rcim.2019.04.005

Zheng, M., Ch. Li, X. Zhang, Zh. Ye, X. Yang, and J. Gu. 2021. The influence of columnar to equiaxed transition on deformation behavior of FeCoCrNiMn high entropy alloy fabricated by laser-based directed energy deposition. *Additive Manufacturing*, 37:101660. https://doi.org/10.1016/j.addma.2020.101660

Zheng, Y., J. Liu, and R. Ahmad. 2020. A cost-driven process planning method for hybrid additive–subtractive remanufacturing. *Journal of Manufacturing Systems* 55:248–263. https://doi.org/10.1016/j.jmsy.2020.03.006

Zhou, H., J. Zhang, D. Yu, P. Feng, Zh. Wu, and W. Cai. 2019. Advances in rotary ultrasonic machining system for hard and brittle materials. *Advances in Mechanical Engineering* 11(12):1–13. https://doi.org/10.1177/1687814019895929

Zhou, Y., and A. Hu. 2011. From Microjoining to Nanojoining. *The Open Surface Science Journal* 3:32–41. https://doi.org/10.2174/1876531901103010032

Zhou, Y., M.H. Hong, J.Y.H. Fuh, L. Lu, B.S. Luk'yanchuk, C.S. Lim, and Z.B. Wang. 2007. Nanopatterning mask fabrication by femtosecond laser irradiation. *Journal of Materials Processing Technology* 192–193:212–217. https://doi.org/10.1016/j.jmatprotec.2007.04.058

Zhu, Y., G. Luo, R. Zhang, P. Cao, Q. Liu, J. Zhang, Y. Sun, J. Li, Q. Shen, and L. Zhang. 2020. Numerical simulation of static mechanical properties of PMMA microcellular foams. *Composites Science and Technology* 192:108110. https://doi.org/10.1016/j.compscitech.2020.108110

Zhu, Y., Q. Zhang, Q. Zhao, and S. To. 2021. The material removal and the nanometric surface characteristics formation mechanism of TiC/Ni cermet in ultra-precision grinding. *International Journal of Refractory Metals and Hard Materials* 96:105494. https://doi.org/10.1016/j.ijrmhm.2021.105494

Zuo, L., X. Zhao, Z. Li, D. Zuo, and H. Wang. 2020. A review of friction stir joining of SiCp/Al composites. *Chinese Journal of Aeronautics* 33(3):792–804. https://doi.org/10.1016/j.cja.2019.07.019

Index

Accuracy, 1, 22, 27, 31–33, 36, 50, 53, 76, 78–81, 87, 107, 124, 131, 136, 143, 152
Additive-subtractive methods, 85, 91, 92, 99
Adhesion, 42, 55, 64, 65, 76, 93, 102, 106, 110, 114, 116
Air, 10, 17–20, 28, 51, 62, 64, 70, 71, 79, 101, 105, 110, 111, 114, 116, 117
Alumina, 35, 38, 50, 67, 103, 104, 110, 124, 142
Aluminate, 64, 105, 142
Aluminide, 66, 109, 110, 144
Aluminum, 8, 21, 22, 32, 39, 45, 51, 52, 55, 60, 66, 68, 73, 92, 93, 105, 108–110, 113, 117, 118, 126, 142, 144, 150
Argon, 28, 51, 60, 65, 114, 119
Artificial intelligence, 127, 128, 131
Automation, 18, 20, 23, 25, 41, 62, 72, 76, 77, 122, 127, 128, 130–132, 134

Binder, 15, 16, 18, 21, 22, 87, 94, 106
Bioactivity, 113
Biocompatibility, 64, 140
Boron, 21, 56
  carbide, 33, 105
  nitride, 38, 106
  oxide, 103
Brass, 51
Brittleness, 16–18, 21, 22, 32–35, 49, 66, 69, 72, 76, 92, 125
Bronze, 51, 142

Carbide
Carbon, 21, 43, 54, 55, 62, 66, 124, 139, 144, 151
  dioxide, 10, 11
  nanotubes, 140
  steel, 117
Cellulose, 6
Ceramics, 1, 2, 6, 22, 35, 36, 38, 43, 44, 49, 50, 52, 53, 55, 64, 65, 67, 87, 90, 92, 102, 104–107, 111, 140, 142, 144
Cermet, 22, 24, 43, 64, 106, 113, 118
Chipping, 33, 66
Chromium, 21, 104–106, 108, 109
  carbide, 115
  nitride, 124
  oxide, 105, 106, 116
Circular economy, 2–4, 74, 85, 100, 135, 152–154
Composite, 1, 2, 8, 24, 39, 51, 64–66, 72, 76, 77, 93, 100, 104, 107, 111, 115–117, 122–124, 142, 144, 150, 151
Contamination, 23, 57, 63, 67, 69, 72, 78, 101, 106
Copper, 1, 24, 35, 51, 52, 58, 67, 98, 108, 114, 117
Corrosion, 9, 29, 50, 53, 63, 64, 77, 101, 109, 117
  protection, 103, 104, 111, 123
  resistance, 1, 6, 41, 43, 53, 65, 74, 76, 83, 94, 104–106, 108–111, 113, 117, 119, 122, 125, 140
Corundum, 21, 67
Cost, 1, 3, 6, 10–12, 15, 21, 23, 24, 26, 41, 46–49, 56, 62–65, 69, 73, 76, 78, 84, 86, 88, 89, 91, 93, 97–100, 105, 113, 117, 118, 122, 123, 129, 131, 136, 139, 142, 145, 148, 151, 152
  saving, 8, 15, 118, 145
Crack, 33, 39, 51, 55, 56, 66, 92, 93, 96, 112, 118, 122, 124
  formation, 33
  initiation, 33, 108
  propagation, 33, 66, 124

Damage, 16, 17, 22, 26, 29, 32, 36, 57, 74, 76–78, 83, 98, 108, 109, 118, 135
Defect, 8, 12, 13, 28, 39, 45, 46, 51, 52, 63, 65, 71, 79, 91, 95–97, 115, 118, 122
Degradation, 2, 4, 5, 9, 83, 84, 93, 109, 151
Demanufacturing, 4
Diagnostics, 20, 24, 51, 140
Diamond, 21–24, 27, 28, 33, 36, 38, 41, 47, 55, 66, 72, 124
Disposal, 3, 6, 73, 122, 151
Ductility, 12, 21, 74, 92, 110
Durability, 8, 43, 64, 80, 81, 107, 116, 117, 123, 125
Dust, 15, 57, 72, 151

Ecology, 100, 154
Elastic (Young's) modulus, 37, 116, 126
Electrical conductivity, 30, 84, 117
Environment, 2, 3, 5, 6, 10, 15, 16, 21, 23, 26, 28, 29, 43, 51, 56, 60, 64, 65, 72, 73, 76, 83, 85, 92, 95, 97, 100, 101, 106, 111–113, 116, 123, 124, 129, 130, 133–137, 139, 140, 151, 152, 153
Epoxy, 8, 77

Failure, 4, 5, 83, 86, 118, 131, 150
  prediction, 24
Fatigue, 5, 33, 36, 37, 55, 92, 93, 96, 110
Filler, 12, 21, 63, 65, 75, 76, 104
Fluoride, 47
Food, 9, 86

Fracture toughness, 37, 55
Functionality, 56, 85, 92, 93, 98, 100, 106
Functionalization, 61

Germanium, 36, 46
Glass, 1, 11, 22, 32, 36, 37, 44, 45, 52, 53, 64, 66, 67, 72, 77, 89, 92, 102–104, 106, 107, 112, 118
coating, 116, 119
fibers, 11
processing, 102
Gold, 2, 44, 140
Graphene, 140
Graphite, 24, 55, 58

Hafnium, 58
carbide, 118
diboride, 111
Hardness, 25, 31, 35, 37, 39, 43, 49, 53–56, 65, 67, 68, 74, 83, 93, 94, 103, 104, 106, 110, 111, 116, 118, 120, 123–126
Heat-affected zone, 33, 39, 50, 53, 57, 63, 75, 76, 150
Helium, 28, 117
Hydrogen, 28, 30, 60, 101

Impact resistance, 12
Inspection, 56, 72, 99, 100
Intermetallic compounds, 56, 92, 102, 105, 144
Intermetallic phases, 76
Intermetallics, 105, 126
Intoxicity, 26
Iodine, 48
Iron, 1, 55, 56, 61, 81, 105, 106, 108, 123

Lifecycle, 1–4, 6, 56, 60, 84, 85, 96–98, 100, 122, 133, 134, 145, 151, 152, 154
Lubricant, 23, 97, 102, 107

Magnesium, 52, 93, 94, 104, 105, 111, 118, 142, 148, 150
Manganese, 61
Material removal rate, 22–24, 31, 36, 37, 62, 68, 72
Measurement, 36, 43, 69, 119, 135, 136, 150
Microhardness, 37, 56, 65, 80
Molybdenum, 21, 22, 105
Monitoring, 51, 52, 75, 86, 133–136, 152

Nanochannel, 42
Nanocomposite, 111, 124
Nanocrystalline, 37, 124, 126, 145, 146
Nanofabrication, 41, 42
Nanojoining, 78
Nanomaterial, 61, 65, 78, 139, 140
Nanoparticles, 44, 74, 139, 140
Nanostructure, 37, 42, 79, 123, 140
Nanotechnology, 73
Nanowall, 42
New materials, 1, 2, 6, 44, 65, 66, 74, 90, 112, 154
Nickel, 21, 22, 49, 51, 52, 56, 91, 104, 106, 113, 117, 143
aluminide, 110, 144
Niobium, 52
Nitrogen, 10, 11, 20, 28, 43, 51, 54, 61, 101, 114, 117, 121
Noise, 15, 72

Oil, 26, 33, 60
Oxidation, 51, 61, 64, 65, 75, 101, 105, 106, 108–111, 113, 119, 120, 124–126
Oxygen, 28, 30, 48, 51, 61, 64, 101, 104, 105, 111, 117, 121

Photopolymer, 87, 90
Plastic deformation, 22, 33, 37, 66, 92, 102, 117, 142, 144, 145
Plasticity, 21, 37, 150
Pollution, 58, 73, 97, 100
Polymer, 1, 2, 6–16, 21, 52, 61, 66, 67, 76, 87, 90, 92, 93, 100, 107, 122, 138–140
Product "end-of-life", 2, 92, 97
Product life cycle, 1–4, 6, 56, 60, 84, 85, 96–98, 100, 122, 133, 134, 145, 151, 152, 154
Protection, 2, 10, 17, 63–65, 83, 84, 101–103, 105–109, 112, 123

Quartz, 72, 103

Reconditioning, 5, 97
Recovery, 2, 4, 5, 32, 73, 74, 96, 100, 123, 152
Recyclability, 6, 85, 88, 100
Recycling, 2, 4, 5, 60, 73, 74, 87, 98–101, 152, 154
Refractory material, 15, 26, 64, 104, 105, 110
Refurbishment, 5
Remanufacturability, 5, 74
Repeatability, 20, 55, 56, 95
Residual stress, 12, 33, 37, 63, 64, 76, 80, 92, 96, 110
Resource, 2–4, 73, 130, 133, 134, 151, 152, 154
consumption, 6, 56, 97, 100, 108
efficiency, 3, 151, 153
savings, 23, 74, 154
Restoration, 2, 5, 22, 57, 83–85, 113, 118, 122, 123, 152
Reuse, 1, 2, 4, 5, 15, 16, 73, 74, 99, 123, 154
Rhenium, 43
Roughness, 13, 15, 22, 23, 31–33, 35–37, 53, 65, 69, 72, 78, 95, 96, 115, 136

Safety, 1, 26, 64, 72, 86, 96, 105, 119, 133, 143
Semiconductor, 41, 44–46, 48, 52

Shape memory alloys, 2, 94, 137, 140
Silicon, 36, 42, 46, 55, 104, 145
  carbide, 21, 67, 68, 73, 106, 115
Silver, 2, 148
Simulation, 10, 15, 22, 23, 52, 119, 128, 134, 135, 145
Smart materials, 1, 95, 137–140, 154
Society, 135, 153
Stability, 62, 64, 80, 84, 102, 105, 107, 110, 111, 120, 124, 135, 140
Steel, 8, 21, 26, 28, 37, 43, 55, 56, 58, 61, 62, 64–67, 74, 76, 77, 81, 93, 101, 105, 106, 108, 117, 144, 150
  alloy, 21, 51, 52, 65, 101, 104, 123
  hardened, 26
  high-speed, 61, 144
  low-alloy, 9, 55, 126
  maraging, 93
  mild, 51, 62, 108
  stainless, 29, 35, 55, 61, 67, 76, 77, 87, 93, 98
Stiffness, 7, 66, 94, 111, 139
Strength, 1, 6–8, 10, 11, 13, 21, 25, 28, 33, 36, 37, 39, 53–56, 62, 65–67, 74, 77, 78, 93, 94, 103, 104, 106, 108, 113, 114, 116, 122, 125, 126, 139, 150
Superalloy, 56, 64, 94, 109, 143, 145
Superplastic deformation, 141–144
Superplasticity, 140–146
Surface integrity, 79, 80, 108

Tensile strength, 37, 66, 93, 94, 96, 98, 141, 150
Thermal conductivity, 21, 26, 38, 53, 64, 76, 93, 94, 104, 106, 111, 112, 114, 115, 118, 120
Thermal expansion, 66, 104, 105, 110, 116, 118
Thermomechanically affected zone, 149
Thin film, 39, 41, 42, 63, 111, 123, 125, 140
Thin-walled parts, 12, 16, 17, 28, 63
Titanium, 21, 35, 45, 51, 61, 87, 89, 91, 93, 105, 118, 144, 148, 150
  alloy, 24, 36, 49, 52, 54, 66, 89, 93, 104, 144
  aluminide, 66
  carbide, 22, 66, 106, 115, 118, 124
  carbide–nitride, 22
  hydroxylapatite, 56
  nitride, 85, 124, 125, 126
Tooling, 8, 86, 93, 98, 99, 135, 136, 142
Toughness, 1, 12, 37, 55, 64, 93, 110, 124
Toxicity, 60
Tungsten, 2, 21, 22, 24, 58, 59, 61–63, 65, 89, 114, 118, 148
  carbide, 30, 35, 40, 43, 123, 124

Vanadium, 21, 60
Viscosity, 8, 10, 11, 12, 26, 35, 102, 103, 107

Wastage, 3, 72, 89
Wastes, 60, 104
Wood, 6, 60

Yttria stabilized zirconia, 64, 112
Yttrium, 43, 45, 110

Zirconia, 64, 67, 104, 112, 142
Zirconium
  diboride, 64, 111, 112
  dioxide, 66
  oxide, 104, 142